THE SACRED
NEURON

EXTRAORDINARY NEW DISCOVERIES
LINKING SCIENCE AND RELIGION

John W. Bowker

I.B. TAURIS

LONDON · NEW YORK

Published in 2005 by I.B.Tauris & Co. Ltd
6 Salem Road, London W2 4BU
175 Fifth Avenue, New York, NY 10010
www.ibtauris.com

ISBN: 1 85043 481 6
EAN: 978 1 85043 481 8

A full CIP record for this book is available from the British Library
A full CIP record for this book is available from the Library of Congress

Library of Congress catalog card: available

Typeset in Goudy Old Style by A. & D. Worthington, Newmarket, Suffolk
Printed and bound in Great Britain by MPG Books Ltd, Bodmin, Cornwall

CONTENTS

For Margaret
with gratitude and deepest love

'I will not always have this hand to hold:
White stone will state the feeling, but be cold.
If this were known, in every waking hour
Tendrils of tenderness would clasp the flower.
Let me then hold this hand for your deep worth
Before we lose our tenure in this earth ...'

PREFACE

Why do we think that things happen in the way they do? Why do we think that some things are beautiful, others ugly? Why do we think that some things are good, others evil? Why do we think that some things are true, others false?

In the twentieth century there were still those who thought that those questions might be answered by reason, in a rational way. Others looked at the failures of reason in so many ways – in morality, for example, in art, in philosophy, even more in the catastrophes of twentieth-century history. As a result, they proposed and they enacted a very different programme, breaking loose from a cultural and political past that had failed. They came to be called 'modernists', and modernism, as they understood it, was a deliberate antagonism against custom and tradition – and even against reason itself. To challenge the confidence of the so-called 'Age of Reason' became a mark of modernity. Postmodernism and deconstruction have taken the destructive challenge even further.

Postmodernism and deconstruction have begun, in their turn, to recede, but not the questions that both they (in our time) and modernism (in the twentieth century) raised about the place of reason. Where do we now stand? Perhaps with the poet Stephen Spender who argued (in *The Struggle of the Modern*) that what is always required is 'the vision of the whole situation'. What is needed, he argued, is a constant integration of undespised experience from the past with the opportunities and novelty of the present.

That is what is attempted in this book. It begins with a person from the past, Hensley Henson, partly because the seeds of this book were sown in lectures that were given under his name, but much more because Henson is a good example of those who hoped that answers might be found to the 'why' questions on which rational people could agree. So I go back to Henson briefly to look at the answers he gave to them. But this is not a book about Henson. My main purpose is to show that while his particular answers are unconvincing, his *trust* in the paramount place of reason is still valid. The answers we now give are very different, but they remain rational, not least because of recent advances in the sciences in general, and in the neurosciences in particular. I am grateful to the electors to the Henson lectures (as also to Hensley Henson) for giving me the chance to attempt a more integrated vision, if not, as Spender hoped, 'of the whole situation', at least of some parts of it that belong together in new and important ways.

Writing a book seemed at one stage impossible. Trouble with my eyes meant that for a time I could not read, let alone do the fundamental work of checking references. Dr Eolene Boyd-Macmillan came to the rescue and did the necessary checking of references with meticulous attention to detail. I am deeply grateful to her. That work was made possible by a generous grant from the Templeton Foundation. The Foundation then made a further grant to assist with publication costs, for which I am equally grateful. Without the Foundation, this book could not possibly have come into being. To John Brooke and Mark Williams I am particularly indebted: given my state of health, I would not have undertaken any writing without their encouragement. The editor, Alex Wright, worked with me on *The Meanings of Death*. It has been good to renew the partnership as he initiates a new religious list with the publisher, I.B. Tauris. To him and to others at I.B. Tauris (especially to Clare Dubois, and to David and Alison Worthington) I owe much. I am grateful also to Quinton Deeley for advice and correction, but even more for his constant support. His pioneering work on the religious brain is just beginning to appear in print. Through the whole process, the support and help of Sarah Brunning have been invaluable. Thanks are due also to Oxford University Press for permission to reproduce material from *The Oxford Dictionary of World Religions*. But above all, my thanks go to Margaret, my wife. In difficult times, she has never faltered. Her comments and criticisms have invariably been right (though almost always, in the first instance, strongly resisted by me). Neither the lectures nor the book could have come into being without her, and to her I offer it as a way, however inadequate, of saying, Thank you.

INTRODUCTION

In October 1854, Lord Cardigan, after a confusion in orders, led the charge of the Light Brigade at Balaklava against the Russian guns. Nearly half of those taking part were killed, evoking from Alfred, Lord Tennyson the famous lines,

> Their's not to make reply,
> Their's not to reason why,
> Their's but to do and die:
> Into the valley of death
> Rode the six hundred.[1]

A hundred years later, Cecil Woodham-Smith wrote *The Reason Why*[2] in an attempt to give the answers: why had the orders been given? Why had they been confused? Why had Lord Cardigan not questioned them when he saw what the consequence would be? Why had the Brigade ridden to certain disaster?

Those are the kind of 'why' questions that historians are likely to ask. In contrast, the science writer Richard Dawkins is clear that 'why' questions should not be asked, at least not by scientists:

> Darwin has shown a sequence of historical events – natural selection – which leads to the existence of all the apparently purposeful things like us. Asking 'Why?' in this case is the same as asking 'How?' That is, I believe, the only explanation – the only question – we're entitled to ask. Suppose that some child is dying of cancer, we say, 'Why is this child dying?' 'What has it done to deserve it?' The answer is, there's no reason why. It's not divine retribution. It's not due to sin. There is no reason other than a series of historical accidents which have led to this child dying of cancer. No reason to ask why.[3]

Their's not to reason why. In other words, human reason must be confined to the kind of blind obedience that sent the Light Brigade to disaster. But obedience to whom? According to Dawkins, it must be obedience to scientists, because they alone appeal to objective evidence with procedures of testing and repeatability that endorse their propositions.

In point of fact, Darwin himself was entirely happy to ask 'why' questions:

How inexplicable are the cases of serial homologies on the ordinary view of creation! Why should the brain be enclosed in a box composed of such numerous and such extraordinarily shaped pieces of bone, apparently representing vertebrae? As Owen has remarked, the benefit derived from the yielding of the separate pieces in the act of parturition by mammals, will by no means explain the same construction in the skulls of birds and reptiles. Why should similar bones have been created to form the wing and the leg of a bat, used as they are for such totally different purposes, namely flying and walking? Why should one crustacean, which has an extremely complex mouth formed of many parts, consequently always have fewer legs; or conversely, those with many legs have simpler mouths? Why should the sepals, petals, stamens, and pistils, in each flower, though fitted for such distinct purposes, be all constructed on the same pattern?[4]

True, Darwin's 'why' questions can be rewritten as 'how' questions: how did it come about that ... ? Nevertheless, they remain 'why' questions. Even more to the point, the physicist Richard Feynman (joint winner of the Nobel Prize for physics in 1965) used to insist to his students that 'why' questions always come first (as they do to young children), because without them, the mind is never stirred up to see a problem:

> Learn by trying to understand simple things in terms of other ideas – always honestly and directly. What keeps the clouds up, why can't I see stars in the daytime, why do colours appear on oily water, what makes the lines on the surface of water being poured from a pitcher, why does a hanging lamp swing back and forth – all the innumerable little things you see all around you. Then when you have learned what an explanation really is, you can then go on to more subtle questions.[5]

Once again, it is possible to rewrite Feynman's 'why' questions as 'how' questions: how does it come about that I can't see stars in the daytime? Even so, they are 'why' questions, and nothing in the reformulation of the question rules out particular answers. 'How did it come about that the child died?' Nothing in the reformulation of the question rules out the answer *as a possibility*: 'because the child sinned; because of divine retribution.' Those answers may seem repugnant, as much to believers as to unbelievers in God. Even then, the question remains with us, *why* do they seem repugnant?

What matters, therefore, in terms of rationality, is not the form of the question, as Dawkins supposed, but the plausibility, or even truth, of the answers; and *that* depends on what is brought forward to support any proposed answer – or to put it in more technical language, it depends on the warrants for our assertions.

That is true for both 'how' and 'why' questions, although there is a real difference between them. There is clearly a difference between asking *how* Jack the Ripper killed his victims, and asking *why* he killed them, between asking *how* the members of the Light Brigade rode their horses and how they died, and asking *why* they rode their horses on that day and why they died. But in both kinds of question, the worth of the answers, in terms of truth and insight, will depend on the warrants put forward to support them.

Dawkins, therefore, would have been on stronger ground if he had written more modestly about degrees of certainty or probability, depending on the warrants that can be brought forward in support of a particular proposition. It then becomes obvious that some questions can be answered with a higher degree of probability than others. 'How many fingers are there on my hand?' can be answered with a far greater degree of certainty than the question, 'How many atoms are there in the universe?'

That, however, means that all questions cannot be answered in the same way, and that the question of what warrants are salient, sufficient and appropriate becomes necessary. Certainly they cannot all be lumped together as though there is only one kind of warrant (let us say, the scientific) for all serious propositions.

It is here that the challenges in recent years to the adequacy of human rationality have been focused. For, while some 'why' questions can be answered with high degrees of probability, others cannot, and only rarely can any of them be answered with absolute and incorrigible certainty.

Thus it is not always easy to say why a particular child died, even after a post mortem. It may be more difficult to diagnose why a child is ill, particularly, for example, with the fast-moving symptoms of meningitis. We move into a different order of difficulty when we try to say why people acted as they did in the past, or why people regard some things as beautiful and others as ugly, or why they judge some things to be right and others wrong, or why a person becomes a suicide bomber, or why I should love my neighbour. In the less well-known *Epilogue to the Charge of the Heavy Brigade*, Tennyson wrote:

> Slav, Teuton, Celt, I count them all
> > My friends and brother souls,
> With all the peoples, great and small,
> > That wheel between the poles.[6]

Why? Why should I feel like that about other people, particularly if natural selection says that only the fittest should survive? As John Turner asked: 'What chimpanzee would sink in its thousands in the mud of Passchendaele for Kaiser, king and country?'[7]

Answers to those questions are, of course, possible, and many have been given. The answer offered by Trivers and Hamilton in 1963 was 'kinship and reciprocal altruism'.[8] The fact that their original formulation has been challenged and changed does not alter the fact that it was an *answer*, and because the authors offered warrants for their assertions, they could be taken, as they were, seriously, but not as though their assertions were beyond correction. Science proceeds, not because it is always certain and complete, but precisely because it is corrigible.

Much the same can clearly be said of the answers proposed to the 'why' questions of history. Historians propose answers to such questions as, 'Why did the Light Brigade make that charge?', and they offer warrants for their assertions by appealing to surviving evidence; but their answers are not incorrigible, and history proceeds, not because it is always certain and complete, but precisely because it is corrigible. Can the same be said of proposed answers to the 'why' questions of aesthetics or of ethics? What, if any, warrants for our assertions can we offer in those areas of human judgement? Or do they belong outside the norms of rationality that obtain in the examples given of science and history? Or even, as some have proposed, outside the norms of rationality altogether? It is the purpose of this book to explore those questions.

This is of course a large agenda for a short book. However, a limit is set on it by the fact that it is based on the Henson Lectures given at Oxford in 2003 – even though it goes far beyond what it was possible to include in those lectures. It is not a book *about* Henson. Nevertheless, it begins with Hensley Henson (1863–1947, bishop of Durham, 1920–39) and remains in conversation with him, because this basic and fundamental question of human rationality (what warrants can we offer for our assertions?) preoccupied him throughout his life, and it is the reason why the lectures that bear his name came into being.

Inevitably therefore he kept on asking where the answers given by faith, or by Christianity, stand in all this. To be dumped, as by Dawkins? Or to be taken seriously because they have warrants to support them? If so, what are they? Or to put the questions in another way, can there be a rational and intelligent faith? That is what mattered to Henson, and that is why all Henson lecturers are required to deal with the topic, 'The Appeal to History as an Integral Part of Christian Apologetics'.

For Henson, as will be seen, this was not an academic question only. He lived at a time when lies and propaganda were rife, and when threats to life and liberty were real. I suppose I must be one of a diminishing number of people who have had hands laid upon them by Hensley Henson - not in confirmation, still less in ordination, but in the kind of blessing that might have been bestowed upon an infant Samuel. I was three years old at the time. It happened, so I was told, on an occasion when Henson was visiting Barking. He had been vicar there from 1888 until 1895, and when my father was ordained, he was first curate of Barking and then vicar of a part of Barking when it was made into a separate parish. Obviously I have no memory of this at all, but my father did tell me years later that Henson had left with him a word of advice. He said (and this was in the year 1938), 'Remember always that there are people at the gates. Your job is to know whether they are the barbarians – or Lazarus.'

The barbarians have not gone away, nor have the poor who lie at our gates. We still have judgements to make and actions to take. Can they, as Henson hoped, be rationally based? Or are they perhaps a matter of opinion, depending on the time and the society in which we happen to live? The ways in which we answer those questions have changed greatly since Henson was alive, but the questions remain as urgent as ever.

The book begins by looking at some examples of human conflict (especially between religions or religious people), and at the way in which Henson hoped that those involved in conflicts of opinion or action might find some neutral ground outside themselves to which their conflicts might be taken for resolution or decision – what I have called 'courts of appeal'. Henson was not alone in that hope, and it seemed to offer the only way in which rationality could enter into human enquiry and argument. Nevertheless, it seems now, in the world of deconstruction and postmodernism, a forlorn hope. I therefore look, in succeeding chapters, at how subjectivity has taken over the ways in which we make many of our judgements, particularly in aesthetics and ethics, but I argue that, in ways very different from anything that Henson could have envisaged, rationality is still paramount. Why that is so is illustrated from current research in the neurosciences which has important light to throw on the ways in which reason and emotion work together in the forming of human judgements. In the final chapters, I look at two practical issues. First, I come back to the involvement of religions (or of religious people) in wars and conflicts, and then I examine the role of religions in the understanding and control of human sexuality.

1 CONFLICT AND THE REASONS WHY

If we opened a newspaper today and read the headline, 'Scandal in Hereford', and if we saw also that there was a bishop involved, we might well think of sex or money or both.

When that headline actually appeared in 1917, it did involve a clergyman, Hensley Henson, who was about to become a bishop, but it had nothing to do with sex or money. It had to do with his beliefs – or rather, his lack of beliefs, according to those who opposed his becoming the bishop of Hereford. How could a man whom his opponents accused of not believing in the virgin birth or the resurrection be consecrated as a bishop? Petitions of protest were signed by hundreds, and angry letters were written to *The Times*.

It is a remote and forgotten conflict. Neither it nor even Hensley Henson is remembered now, beyond perhaps his famous remark to Lang, when Lang was archbishop of York. Sir William Orpen had just painted the official portrait of Lang who stood before it and lamented to Henson, 'They say in that portrait that I look proud, prelatical and pompous', to which Henson replied, 'And may I ask Your Grace to which of these epithets Your Grace takes exception?'[1]

And yet, although the details of that conflict in 1917 are now forgotten, the underlying issue remains, the issue of what people mean by what they say that they believe. I believe in relativity, in the United Nations, in hot whisky and orange for a cold. But what do I mean by what I say that I believe? And what warrants do I offer in support of my beliefs? Margaret Fuller once came up to Thomas Carlyle (that austere Scot whom Elizabeth Barrett Browning heard 'growling about the Crystal Palace of the Great Exhibition' in 1851, saying that there is 'confusion enough in the universe without building a crystal palace to represent it'[2]) and said to him effusively, 'I believe in the universe', to which Carlyle replied, 'By God, she'd better!'[3] What did she on the one hand and Carlyle on the other mean by 'believing in the universe'? Henson had no problem about standing up in church and reciting the words of the creed, but did he do so, as one of his opponents asked, 'in good faith'?

Henson had no doubt that he did, and by 'good faith' he meant, not the mere recitation of words, but the intelligent understanding of the words to which a person puts the signature of 'I believe'. That is the issue which remains always with us. When David Jenkins was to be consecrated as bishop of Durham in 1984, his opponents accused him of dismissing traditional beliefs in the resurrection as a juggling trick with a bag of bones, and when York Minster was struck by lightning, there were some among them who regarded this as an act of God, a sign of God's anger against the liberal erosion of faith – and the words 'liberal' and 'liberal agenda' have become, in such conflicts, words of venom and abuse.

But what Jenkins in fact was saying followed very much in the steps of Henson, his predecessor at Durham: any account of the resurrection that makes it look as though God was doing a juggling trick with bones is far removed from the first accounts of the resurrection in the New Testament.[4]

As with Jenkins, so with Henson: in conflicts that throw up such colossal clouds of dust, where is truth to be found? To what can we appeal when we are divided so deeply that the lines of schism seem inevitable? When Gene Robinson was elected bishop of New Hampshire in 2003, to some this seemed a contradiction of the plain Biblical prohibition of same-gender sexual acts, to others the Bible has to be related to those new understandings of homosexuality that the Bible could not possibly envisage. To some, that is subordinating the Bible to the modern (and liberal) agenda, to others it is recognizing that the world set the agenda for the Incarnation, and that always the consequence of God in Christ is extended by the Spirit into that same world as it is better and differently understood.

No doubt to some the argument seems trivial. Who cares any longer what bishops do or do not believe? As with the famous remark attributed to Robert Huttenback when Chancellor of the University of California ('The reason why academic disputes are so vicious is because the stakes are so low'), so here: in the opinion of many, religious arguments are a matter of the unbelievable being upheld by the unprovable – an opinion often summarized in the form that these are arguments about the number of angels that can dance on the point of a pin. So why not let those who have a fancy for such things get on with living in the past, while we get on with living in the twenty-first century?

An immediate answer is of course that people with religious beliefs are also living in the twenty-first century, where some of them have made their beliefs a reason or a justification for acts of violence and terror. At the very least, we need to understand why people hold their beliefs so strongly that they are prepared to die and to kill for them. For more than 40 years I have been writing on religions, warfare and violence, explaining why religions (or more accurately, some people with religious beliefs) are likely to destroy human life as we know it on this planet.[5]

At the heart of it lies what I have called 'the paradox of religious urgency': religions are only such bad news because they are such good news. Religions historically have protected and transmitted from one generation to another almost all the most important human achievements - in the exploration, for example, of our own nature and its spiritual competence, in art, music, agriculture, poetry, dance, drama, above all in the ways to procreate another generation and bring it up successfully (this vital point is explored in Chapter 6). They have protected and transmitted the ways in which humans can come to know and be known by God - even while their understandings of God change, just as their understandings of the universe change. Even the natural sciences have their origins and roots in religion. It is only very recently, in the last two or three centuries, that the two have come apart. The very issue now used to deride the irrelevance of religious arguments, about the number of angels that can dance on the point of a pin, was in fact a crucial moment in the history of mathematics, leading to the debate about potential or actual infinities, an issue that was only resolved by Georg Cantor at the end of the nineteenth century.[6]

It is because religions have so much of such great importance to protect and transmit to other people and to successive generations that they have developed as organized systems to ensure that the protection and transmission happen. But systems need boundaries to preserve them, and boundaries, whether literal or metaphorical, all too often lead to border incidents when those boundaries are thought to be under threat. That is one reason why religious wars are common, and why virtually all the long-running and apparently insoluble problems in the world (for example, in Northern Ireland, Cyprus, the Middle East, Sudan, Kashmir, Khalistan, Sri Lanka, India and Pakistan) have deep religious roots. If you wish to see where future conflicts will occur, draw on a map of the world the lines where religions or subsystems of religions meet - as, for example, in Nigeria. When the boundaries are metaphorical (for example, the incursions of secularism on ways of life and belief that are authorized in a particular system), the dangers are equally great, because 'the enemy' might be anywhere in the world. The implications for terrorism are obvious.[7]

It is true that religions are capable of coexisting with each other, and often have done so for long periods of time (that has been the case, for example, until very recently, with Jewish communities in Muslim lands).

But they do so only as long as the continuity of each of the religions in question does not seem to be threatened. When threats do develop, there are always people who will die rather than abandon the inherited treasure, especially when they themselves have tested it and proved it to be of worth in their own case.

This is the paradox of religious urgency: religions are such bad news (when they are) only because they are such good news: they protect so much that is so important and so well tested through time that people would die rather than lose it.

This is not in any way an argument that religions are *always* good news. Far from it: it is easy to see the great harm that has been done in religions, by way, for example, of spiritual terrorization, or of the subordination of women in most aspects of their lives to the decisions and determinations of men. Nor is it to argue that religious beliefs *alone* are the cause of those and other wars. The causes of conflict are always many and varied, and some belong only to each particular conflict. But among them, very often, the appeal to religious belief and religious emotion is a powerful warrant for action.

That perhaps seems obvious, but the underlying point is important. In giving the reasons why something has happened, we usually search for an explanation in terms of the cause or causes that seem to us most immediate and direct. As Samuel Butler observed, 'people generally become fatigued after they have heard the answer to two or three "whys" and are glad enough to let the matter drop.'[8] Yet in fact the reasons why things happen are never simple, although we are inclined to make them so. Our inclination (life being short) is to mention only the most obvious causes, or the ones that seem particularly relevant to us: why (how) did it come about that this child is dying (pp ix–xi)? Because of the cancer. Yet for all complex events and eventualities, the reasons why they come into being are always many. Why in general does cancer occur? Why *this* child and *this* cancer? Why do some die and others not? As Coveney and Highfield make the point strongly in their book on the unfolding complexity of the universe, the reasons why things happen are usually so many and so varied that they *cannot* be reduced to single or simple causes:

> Many people have accepted the reductionist message of contemporary science. Although sometimes powerful, reductionism can be destructively simplistic. If a mother loses her son to cancer, she desperately hunts for an explanation: was it that artificial coloring in his favorite orange juice? Was it the electric power cable outside his bedroom? Was it the cigarette smoke he inhaled? Sometimes, a simple cause can be established, but often one cannot be found. ... Complexity teaches us that effects can have an irreducible tangle of causes.

Just as the properties of a cement slurry depend on a vast number of contributing factors, so, too, does the state of our health.[9]

Clearly, it is impossible to list *all* the reasons why something has happened, and as a result we have to choose what we are going to include, and also what we are going to exclude, by way of explanation. Thus while religions alone are not the cause of the kind of conflicts mentioned above, it is extremely rare for politicians to *include* religious commitments among the causes. The recent work of Johnston and others[10] is exemplary in the way they show the salience of causes derived from religions, not just in general, but in particular conflicts. They make the claim, not that religions are the sole cause of conflict, but rather that 'dogmatic secularism' or 'secularizing reductivism'[11] prevents politicians and others from recognizing the importance of causes derived from religious considerations, both in conflict itself, and also in seeking peace and reconciliation. As I put it more despairingly of the political scene in 1982, 20 years before the invasion of Iraq:

> One of the most obvious reasons why we seem to drift from one disastrous ineptitude to another is, ironically, that far too few politicians have read Religious Studies at a University. As a result, they literally *do not know* what they are talking about on almost any of the major international issues. They simply cannot. It is time we began to educate ourselves, not just in economics, or in politics, or in technology, but also in the dynamics of religious belief and continuity, because whether we like it or not, it is religion which still matters more than anything else to most people alive today.[12]

As with Johnston, so here: this is not a claim that religion is the sole cause of conflict in those instances where religions are involved. Indeed, it is important to bear in mind that nothing ever is 'the sole cause' of anything. That is why I have argued for many years that if we want to find out the reasons why anything has come into being as it has, or why anything has happened as it has, it is far wiser not to seek the cause but to specify the constraints. The point about constraint is that it *includes* active causes of a direct or indirect kind, but it also alerts us to a far wider network of what has brought an eventuality into being, including passive and domain constraints. If I push a book, the fact that the book moves is indeed because I pushed it, but also because both I and the book are constrained by (amongst much else) the laws of motion.

Of course in the ordinary business of life, and especially in the business of offering scientific explanations, we do not have time to specify *all* the constraints that have controlled an eventuality into its outcome, into its being what it is. Therefore in explaining any phenomenon, we choose from the whole range of actual constraints those that relate most closely and immediately to our concern, and leave the others as an unspoken assumption. In making those choices, what is known in artificial intelligence research as 'the frame problem' is a paramount necessity.[13] But the fact remains that we do have to choose. If, as loss adjusters for an insurance firm, we ask, 'What caused that fire?', we are unlikely to specify, 'the presence of oxygen'. Yet, if we are seeking to explain the outbreak of fire in an unmanned space capsule, we will undoubtedly want to include the presence of oxygen in the specification of constraints.

So how do we choose? It is at this point that a version of Ockham's razor is usually wielded ('where one explanation will do, don't multiply explanations'), but Ockham's razor has virtue only so long as you do not use it to cut off your own head. As I put it in *Is God a Virus?*:

> Where additional constraints *must* be specified in order to account for an eventuality, nothing is gained by insisting, in the name of Ockham, on only one. A better principle is this: be sufficiently, but not recklessly, generous in the specification of constraints; or at least otherwise be modest in what you claim to be 'the true and only explanation'.[14]

Even so, the fact remains that nothing dictates or controls which reasons, or which sets of reasons, we choose. Some, for sure, will seem irrelevant, frivolous or absurd. Equally, 'choice' is often the wrong word, since we make use of certain reasons without giving the matter much or indeed any thought. In other words, we operate our modes of explanation from well-tried precommitments which sometimes take the form of prejudice – as, for example, when we state that scientists do not ask 'why' questions (p ix). As Miller summarizes the point:

> When is a description of causes helping to bring about something informative and thorough enough to explain its occurrence? No general answer seems right – and this is the first step toward the right answer. ... In some cases, a standard requiring a list of causes sufficient under the circumstances is too weak. But in other cases, a standard requiring a list of causes sufficient in themselves is too strong. We need a compromise between these two extremes.[15]

That is why, in giving the reasons why something has happened, it is not possible to rule out God in the total specification of the constraints. We may not choose to do so (we may even regard such a suggestion as 'irrelevant, frivolous or absurd'), or we may prefer to speak of God as supplying the condition of constraint as a consequence of which all things exist.[16]

Even so, the constraint of God is clearly actualized in the agency of human beings at least, and it is the reason why prayer for others is consequential without violating 'the laws' of the universe – i.e. those other constraints which bring particular eventualities into being, and which, some would argue, seem to be far more accessible than God. Even if that were so, it is still a mark of inhumanity to try to put a stop to the question, 'Why?' When Primo Levi was held overnight by the Nazis in Buna concentration camp while being transported to Auschwitz, he left indelibly inscribed on every moral memory the reason why the question 'why?' is so important:

> Driven by thirst, I eyed a fine icicle outside the window, within a hand's reach. I opened the window and broke off the icicle but at once a large heavy guard prowling outside brutally snatched it away from me. '*Warum?*' I asked him in my poor German. '*Hier ist kein warum*' (there is no why here) he replied, pushing me inside with a shove.[17]

Henson resisted fiercely the brutality of human conduct in the rise of totalitarian regimes, an evil which seems so often to be past all understanding. But understanding the reasons why such things happen is a part of the resistance to them. Consistently, Henson also believed that rationality is needed in the other, albeit lesser, conflicts in which he found himself – as during the uproar that arose when his name was put forward to become the bishop of Hereford. Maybe the particular issues and arguments seem as remote to us as the arguments of the early Church seemed remote to Uncle Toby when he dismissed them as 'a pudder and a racket about words of little meaning and as indeterminate a sense.'[18] But the questions still are fundamental:

- To what do people with competing ideas appeal to support or justify what they are saying?
- What warrants do they offer for their assertions?
- Are those warrants sufficient or adequate for the task?

Those are the questions of rationality, and they are not confined to religion. They are equally important in science. When, for example, Alfred Wegener argued in *The Origins of Continents and Oceans* (1915) that the two continents on each side of the Atlantic had once been joined together and had drifted apart, he thought that he had produced sufficient warrants for his assertion to refute the existing theory that they had been joined by a land bridge that had subsequently collapsed. His colleagues were not convinced,[19] and it was not until decades after his death in 1930 that the discovery of a ridge running down the middle of the Atlantic supplied the final warrant for his basic assertion – even then, his idea of continental drift was amended to the theory of tectonic plates, to which the evidence now pointed.

That example shows that this issue goes far beyond religion and religions. It is fundamental to human reason and rationality. What warrant do I offer to justify my assertions? Warrants alone may not provide conclusive demonstration or proof. They may establish only probability. That is why most scientific propositions are approximate, provisional, corrigible, and often wrong from the point of view of later generations. Wegener's proposition about continental drift had to be corrected by those who came after him, even though his basic insight was right. The proposition about reciprocal and kinship altruism put forward by Trivers and Hamilton (p xii) had to be corrected when it became clear that kinship is understood by many humans to apply as a metaphor to the entire human 'family', and that in any case the key developmental interactants maintaining adaptive social behaviours are not confined to genes. Even so, their basic insight was important.

If, then, many of the judgements of science are corrigible, so too are the judgements put forward by historians. To give an example, in 1915 the Misses Dodds gave an account of the Pilgrimage of Grace (1536–37) which Geoffrey Elton described as 'this magisterial discussion [which] held the field quite undisputed for some fifty years.'[20] They offered conservative and religious reasons why the Pilgrimage had taken place.[21] But these were challenged and changed from the 1960s onward, despite the fact that the evidence on which they relied was not disputed.[22]

In the case of science and of history, it is easy to see the different ways in which warrants are offered in support of assertions, and the reason why those judgements (again, in different ways) are corrigible. But such judgements are not a matter of private or subjective opinion only. Does the same apply to the judgements we make in aesthetics and ethics – the claim, for example, that the current winner of the Turner Prize has produced a work of aesthetic excellence, or the claim that abortion is always morally wrong?

At first sight it would seem that these *are* matters of private opinion because it is difficult to produce warrants for assertions in these areas that others can, or indeed must, accept. People are deeply, even passionately, divided over the Turner Prize or over abortion. Even so, they do offer reasons why they are making the judgements that they do. What is the status of those reasons? Are any of them equivalent to the warrants offered on the basis of evidence in the case of history or of science? If they are to advance beyond being matters of opinion and if they are to be more widely accepted, will they not require something like a warrant for their assertion, even though such warrants do not lead to conclusive demonstration or proof? The judgements may still be contested (as they clearly are), but on what grounds might one aesthetic or ethical judgement seem more probable than another?

All these are fundamental questions of rationality, and Henson believed that religious claims cannot evade them. In contrast, those who opposed him argued not only that they *can* be evaded but that they *ought* to be evaded, because religious assertions rest on faith, not on reasons and warrants: Paul wrote that the life he lived he lived by faith,[23] Krishna declared, 'Give up all activities [lit. *dharma*] and come to me alone for refuge,'[24] God through Muhammad proclaimed, 'Whoever puts his faith in God, that is sufficient for him.'[25] Faith that looks for warrants is surely no longer faith: it has become the conclusion of an argument, reducing God to the measure of human rationality. In any case, it has become a commonplace to argue that religions have less to do with truth than with temperament, with the feelings people have and the stories they construct about themselves, about the world, and about their immediate and ultimate prospects.

Yet paradoxically faith itself turns into the warrant offered for the assertions that rest upon it – that Jesus died for my sins, that Vishnu became incarnate in Krishna to deal with disorder in the world, that God revealed the Quran through Muhammad. But the appeal to faith alone will resolve neither the conflicts that arise between religions nor the conflicts that arise within religions.

So the question for Henson remained whether there are any courts of appeal to which two sides in conflict might take their case for adjudication. The most obvious in the case of religions must seem the court of scripture or of revelation. If humans have received a direct revelation from God, then that surely moves the game from probability to certainty. But what is the warrant for the assertion that the Vedas, or the Tiruvaymoli, or the Guru Granth Sahib, or Tanakh, or the New Testament, or the Quran, to mention only a few, is that revelation? Perhaps they *all* are. If it is right (as it clearly is) to use the phrase 'concursive writing'[26] to acknowledge the distinctive phenomenon of the Bible and to link it to the word 'inspiration', then it will be equally right to use it of those other works (though Muslims and some Hindus would dislike it).

But that does not resolve conflict between them or within them, as Joseph Wolff found out when he went as a missionary to Arabia:

Wherever he went, Wolff argued. He argued with Christians and Jews, with Hindus and Muslims, with Catholics and Protestants, with Sunnis and Shias. He argued about almost everything: about the Pope and the Millennium, and Mohammed, and the Lost Tribes, and the Second Coming, and the end of the World, and about what would happen to all the fishes when the sea dried up. He argued good-humouredly, tirelessly, and without any regard whatever for the consequences. ... On his way through Arabia he engaged in a series of theological discussions, which, even by his standards, were unusually lively. One of these was with a dervish, who, on being asked politely whence he came, replied: 'Dust is my native land and to dust I shall return. Ho! Ho! HO!', the last 'Ho!' being 'uttered in such a powerful voice that it produced an echo'; and yet another with some Wahabites to whom he had given some copies of the New Testament. 'The books you gave us', said the Wahabites severely, 'do not contain the name of Mohammed, the Prophet of God.' 'This circumstance', Wolff rejoined, 'should bring you to some decision.' 'We *have* come to a decision!' replied the Wahabites, with fury stamped upon their faces, and having horsewhipped Wolff tremendously, went about their business.[27]

The truth is that virtually all human judgements of any interest are open to contest. What Henson hoped was that he could identify certain 'courts of appeal' to which these contested judgements might be taken for some further adjudication, perhaps even for arbitration. Intelligent rationality may not produce conclusions with which all must agree, but it can exhibit why some proposals are more probable than others, and why some are so drastically improbable that they can be regarded as just plain wrong. What Henson sought was intelligent rationality as much in religious belief as in history or science or aesthetics or ethics.

That is why the Henson Lectures in Oxford must be given on the theme: 'The Appeal to History as an Integral Part of Christian Apologetic'. History was for Henson one of these 'courts of appeal', to which might be taken such conflicts as those that assailed him when he was proposed as the new bishop of Hereford. There were other 'courts of appeal', as we will see.

But was he not completely wrong? Henson was a true child of the Enlightenment in his belief that the world and Christian faith within it can be intelligently and rationally understood. But in recent years a chorus of voices, ranging from a conservative Pope to radical postmodernists, have been assuring us that the Enlightenment programme was a disastrous mistake. According to John Paul II, 'a *forma mentis* (mind-set) born of the Enlightenment – not only the French Enlightenment but the English and German as well' banished God from the world: 'The rationalism of the Enlightenment put to one side the true God – in particular, God the

Redeemer. The consequence was that man was supposed to live by his reason alone, as if God did not exist.'[28]

Postmodernists are equally scathing about the Enlightenment project of rationality:

> The God of the traditional philosophy of religion is a philosopher's God explicating a philosopher's faith, to be found, if anywhere, only on the pages of philosophy journals, not in the hearts of believers or the practice of faith. This philosopher's God is a creature of scholastic, modernist, and Enlightenment modes of thinking that deserve nothing so much as a decent burial.[29]

The Pope and postmodernists do not otherwise have much in common, but what they both say here is very simple: Reason is not sufficient to deliver us from evil.

That in itself is true. But Henson stood very firmly with the conclusion of François Jacob's book, *The Possible and the Actual*:

> The enlightenment and the nineteenth century had the folly to consider reason to be not merely necessary but even sufficient for the solution of all problems. Today, it would be still more foolish to decide, as some would like, that because reason is not sufficient, it is not necessary either.[30]

Even so, Henson's hope that he could find some independent court of appeal to adjudicate on conflicts of judgement, whether in belief or in ethics or in aesthetics, seems, in these postmodernist days, to be a particularly forlorn one. That is why I gave my own Henson Lectures under the title, 'Did Henson waste his money?' So: did he? Is it possible, as he hoped, to hold a rationally intelligent faith connected with the worlds of science, aesthetics and ethics? Henson opted for reason. It did not make for an easy life.

2 THE APPEAL
TO HISTORY

I have called Henson 'a child of the Enlightenment' (p 10), and so he was, at least chronologically. He was born in 1863 and died in 1947. But that meant that he lived through the whole period which demonstrated how completely incapable reason is to deliver us from evil, and Henson was well aware of that. He lived through two world wars, and he saw the rise of the dictators – and about them he had no illusions. Henson always spoke his mind clearly and often aggressively, and on the dictators he spoke and wrote with consistent opposition. For example, in 1936, he wrote this letter just before he was due to meet von Ribbentrop, the German ambassador:

> Tonight I am going to Wynyard to meet the German Ambassador, Herr von Ribbentrop, who is making a private visit to Lord Londonderry: and I am wondering how I shall succeed in maintaining courtesy without concealing the moral nausea which the Hitlerite regime creates in my mind. There seems a growing expectation that war on the grand scale will shortly break out, and for that supreme disaster, which may well make an end of what we call Civilization, Hitler will be chiefly responsible. He divides the credit with his fellow Dictator – Mussolini. The crime which has destroyed the last vestige of liberty in Abyssinia has opened the door to a whole flood of disasters. I think you saw the pamphlet which I wrote to relieve my own mind. It can be, alas, no more than a wreath placed on the coffin of African liberty. The effect of this successful villainy cannot but be far-reaching.[1]

'The pamphlet' to which he refers was written earlier in 1936 in protest against what he called 'the rape of Abyssinia by Mussolini',[2] and he sent a copy to every member of the House of Commons – even though, as he realized, it would make no difference to events. But he did not give up. He was by then bishop of Durham with a seat in the House of Lords, where he made his voice of protest heard. Indeed, his last speech in the House of Lords, in 1938, was on this very subject, although he recognized how unpopular it would be. In his *Retrospect*[3] he wrote:

> I shall make myself very odious if I criticize the Government's foreign policy, but I mean to do so, if I get the opportunity. It can do nothing but good to raise a protest, even though one knows in advance that nothing effective can

come of it. After all, one owes something to one's own self-respect, and, as a spiritual peer, I ought to speak on an issue which does assuredly raise the question of moral obligation.[4]

How right he was in his prediction! He made, on his own account, a poor speech (he called it 'a complete failure'), whereas he thought that Lord Halifax[5] made a good one, after which, he wrote, 'a good many peers inserted knives into that poor gentleman [the bishop of Durham]', but as he had already left the chamber, he could not, as he put it, 'appraise the measure of injury he had received'.[6]

Henson knew very well how extensively propaganda and the means of its dissemination (including what we would now call 'spin') were being used. The Italian claims to Abyssinia were only one example among many of the deliberate 'rewriting' of history in order to tell persuasive stories. It was a major reason why Henson believed that history itself must serve as a court of appeal for those concerned with truth. When such misuse is made of history, it is only to history that appeal can be made – that is, to history more dispassionately researched and understood.

It is easy, therefore, to see why Henson regarded history as so important, and why the appeal to history played such a fundamental part in his own life. He himself had read history at Oxford from 1881 to 1884. He was then elected a Fellow of All Souls, and in 1908 was invited to become the Regius Professor of Ecclesiastical History. He refused the invitation on the ground that an immense change had taken place in historical studies. As he wrote to the archbishop informing him of his refusal: 'The literary and pictorial historian has been replaced by the patient and infinitely laborious historian.'[7]

What Henson understood by 'a literary and pictorial historian' was exemplified for him by Carlyle. In his view, Carlyle wrote his account of *The French Revolution* in the style of a nineteenth-century novel. Henson's contemporary G.V. Routh once described Carlyle's book as 'history after the manner of Victor Hugo', in which 'individuality is not lost in crowd psychology, much less in the "tendencies of the age".'[8]

With that last point, Henson agreed, because it makes history a story in which individuals and not ideologies (whether Marxist or Fascist) are the agents of events. The meaning of history in *that* context is one in which individuals march out to make themselves and the world better – or indeed, all too often, worse. Henson marked with approval this passage from Carlyle's *Past and Present*:

Not a May-game is this man's life; but a battle and a march, a warfare with principalities and powers. No idle promenade through fragrant orange groves and green flowery spaces, waited on by the choral Muses and the rosy Hours: it is a stern pilgrimage through burning sandy solitudes, through regions of thick-ribbed ice.

Whether the Bishop's Palace at Durham is best described as a burning sandy solitude is doubtful, even though bedrooms in such palaces could indeed be regions of thick-ribbed ice. But 'a battle and a march' Henson understood extremely well, because they were constant themes in his own life, not least in support of individuals against ideologies. All ideologies, whether secular or religious, were to him suspect, unless they conformed to the test of truth. For him, any true story, not least concerning the past, must be well grounded in evidence, in what he was more inclined to call 'facts'. That is what he meant by the move from the literary to the patient and infinitely laborious historian.

Henson did not regret that change in the style of history, because he had already come to the conclusion that it was wrong to follow Hegel in producing grand theories of the story that history tells – the self-development, as Hegelians were inclined to say, of the Absolute Idea as it becomes manifest in the process of time. It was not an inevitable conclusion for Henson to have come to when one remembers that T.H. Green was still alive when Henson arrived at Oxford to read history – Green did not die until 1882, and Henson had some sympathy for Green's *political* liberalism. But where history is concerned, Henson was clear: the appeal to *some* facts will always illustrate a story, the appeal to *all* facts (or at least to all available facts) may call that story into question.

That is why he refused to discuss, let alone countenance, the claims of people who took no account of history. In 1933, he responded to someone who had asked him about the British Israelites. He replied in his characteristically pugnacious style:

> The British-Israelite Movement in my judgement lies below the level on which rational discussion is possible. It belongs to the numerous class of fatuities which pullulate so rankly throughout the English-speaking world, and are all conditioned by two cardinal errors, viz: Fundamentalism in handling the English Bible, and a complete failure to understand historic Christianity. On the soil of these twin and connected misunderstandings there grows a queer amalgam of British Imperialism, Protestant intolerance, and individualistic pietism, which has a strange attraction for the half-educated. ... Argument is absolutely futile.[9]

So Henson never lost his conviction that the work of historians cannot be simply to select those facts which exemplify or illustrate the larger story that they are determined in any case to tell: in contrast, the more professional historians become, and the more their work is well founded in evidence, the more useful and decisive history becomes. But useful in what way?

That question is the key to understanding Henson's 'appeal to history'. Useful in what way? To get Henson's answer to that, we have to travel back to the year 1917. It was a year that Henson called 'stormy'. That at least is the name he gave to Chapter 5 of his autobiography, 'A Stormy Year, 1917'. Why? Because in that year there were two major uproars in which Henson was involved.

The first was the forming of the group that called itself 'Life and Liberty'. Life and Liberty was a group that came together during the First World War in order to secure for the Church of England greater freedom in making its own decisions – freedom, that is, from Parliament. It was the success of this group that led first to the Church Assembly and then to the present General Synod. The group won its way mainly through the energy of William Temple,[10] but the leadership and certainly the inspiration came from Charles Gore, who at that time was still bishop of Oxford.[11]

Henson was emphatically opposed, partly because he believed that this would be the first step toward disestablishment – and he objected to that because he liked the way in which Parliament gave a voice to the entire lay population of the country, and not primarily to the clergy (he knew himself and past history well enough to know that the clergy cannot be trusted with the running of the Church on their own); and he objected also because he regarded Gore as the head of a Church party that was trying to impose its views – in this case, Anglo-Catholic views – on the Church at large. To Henson, this was a clear example of a grand narrative being imposed on the historical evidence.

So at the inaugural meeting of Life and Liberty in 1917, Henson voted against the resolution that the two archbishops should find out 'whether and on what terms Parliament is prepared to give freedom to the Church in the sense of full power to manage its own life'.[12] When the vote was counted, he found himself in a minority of one. 'The meeting,' he commented later, 'was "Gore's crowd". In it I was the solitary dissentient.'[13]

In his *Retrospect*, he immediately went on to explain why, as he put it, 'Gore and I were once more publicly opposed.' It came down to history as the proper court of appeal:

In Gore's view of Christianity the form of Church polity, Episcopacy, had essential importance; in mine, no form of ecclesiastical polity could ever be more than a non-essential, varying from age to age, always the creature of time and place. He was ever appealing to 'principle', I to history. Obviously, on the

specific issue of Church and State, our difference could not but be profound.[14]

It is this appeal to history that Henson took from his time at Oxford, and
it remained paramount throughout the rest of his life. In the year before
he died, he wrote to the dean of Winchester applying the test of history to
the virgin birth and the resurrection:

> I have ever contended, and do still contend, that an historical fact is a fact cer-
> tified to be such by adequate historical testimony, and that neither of these
> two so-called historical facts, viz. the Virgin Birth and the physical Resurrec-
> tion, is in my opinion, and in that of many historical students, whose belief in
> the Incarnation cannot be challenged, adequate.[15]

Thirty years earlier, in that same stormy year of 1917, those two items in
the creed, the virgin birth and the resurrection, had been exactly the ones
that had again aroused the fury of Gore when it was proposed that Henson should be made bishop of Hereford. They had clashed over Life and
Liberty – the first major uproar of that year.[16] They clashed again in the
second uproar, the proposal to consecrate Henson as a bishop. Did Henson believe each of those two clauses in the creed?

Since Gore was clear that Henson did not, Gore made a formal protest
to the archbishop against Henson's consecration which was duly printed
in *The Church Times* under the headline, 'The Hereford Scandal'. An
explosion of hostility sent into battle against him those whom, as Henson
put it, he had offended on other grounds. He listed them:

> Feminists, Sabbatarians, Total Abstainers, Fundamentalists, Spiritualists,
> 'Corybantic Christians' of many descriptions, 'Ritualists' of varying degrees of
> absurdity, Faith-healers, the fire-eaters who clamoured for 'Reprisals' against
> the Germans, and the inexorable patriots who refused to distinguish between
> genuinely conscientious Pacifists, and mere shirkers – all these had reason for
> regarding me with disfavour, and naturally were not unwilling to see me im-
> mersed in adversity.[17]

The issue, as Henson saw it, was, once again, history and historical judgement, and the way they should be applied to the Bible. Over the previous
50 years, immense battles had been fought over Old Testament criticism.
Henson insisted that the same criteria of historical criticism must equally
be applied to the New. Those were the grounds for his objections to
Michael Ramsey's[18] book on the resurrection, as he wrote in a letter:

> It confirmed a certain misgiving about him, which his inaugural lecture
> (which he very kindly sent me) stirred in my mind. Does he really understand
> historical method? Does he imagine that History can be seriously handled un-
> der the domination of theological assumptions? To my mind Theology cannot

begin its work until History has provided so much of its material as is properly called 'historical'. The weakness of 'orthodoxy' modern and ancient has ever been its habit of subordinating History to Theology, for the Religion of the Incarnation is bound into the historical process, and cannot be safely 'developed' except on a sound basis of historical science. It is the historian's <u>primary obligation to cleanse his mind from prejudice</u>, whether friendly or hostile, to the orthodox theory; but 'prejudice' does not mean the natural and unavoidable bias or tendency of the student's mind; it does mean his subordination of the historical argument to the requirements of a theory. But I must stop; or I shall weary my eyes, and spoil your admirable temper![19]

Always for Henson the court of appeal is history and historical fact, whether in theology or in confronting the grand narratives of the 'isms' of his day: fascism, socialism, Marxism, totalitarianism – or, in the case of the Bible, fundamentalism.

Where the Bible was concerned, Henson did have one important ally in that stormy year of 1917. At the height of the storm, an Oxford professor wrote to *The Times*, and his letter carried much weight. The letter was from the Lady Margaret Professor of Divinity at Oxford, William Sanday, of whom *The Oxford Dictionary of the Christian Church* says, 'His influence went far in winning Anglican clergy to the acceptance of modern methods of NT study.'

Sanday, like Henson, espoused historical fact as the foundation of all else, certainly of theology. In his Cambridge lectures he claimed: 'That is really the meaning of all Christian theology. The facts come first.'[20] So he argued that Christian theology cannot be divorced from Jesus, and that consequently people cannot embark on theology until they have first secured the historical facts of the life and ministry of Jesus – hence the strong emphasis on Biblical history in the theology syllabus at Oxford University until well after the Second World War. Like Henson (p 13), he believed that this required hard and laborious work:

> One may be right by accident, or through acquiescing in someone else's opinion that happens to be right. But to deserve to be right is another matter. For that, the conditions are exacting and severe.[21]

The trouble is that by the time people in each generation have sorted out the historical Jesus, they are already dead or retired. That is why, according to Sanday, Anglicanism produces great Biblical scholars and poor theologians. Incidentally, he paid the price himself, because, as *The Oxford Dictionary* also puts it, 'His long-projected Life of Christ was never finished.'[22]

All this explains why 'the appeal to history as an integral part of Christian apologetic' is so precisely specified as the topic for the Henson lectures. In 1917, the storm had broken about Henson and he had weathered it by putting his trust in God and in the appeal to history and to historical fact as he understood it. But whatever that may have meant to Henson when he died in 1947, and however straightforward 'the appeal to history' may have seemed to him then, it cannot be so now. Henson accepted that all historians are subject to 'a natural and unavoidable bias'. What he did not appreciate was the extent to which that makes all historical judgement *relative* to the historian. The fuse of cultural relativity had already been lit and was shortly to blow up Henson's hope of an appeal to history as he understood it.

Cultural Relativity

Cultural differences are so literally obvious that a person only has to take a look to see them. What have varied and changed have been the ways in which cultural differences are understood and interpreted. The simplest conclusion, at one extreme, has been to say that other cultures are inferior to one's own, and to act on that basis: the Japanese expelled and excluded foreigners for two centuries on the grounds that, as Hirata Atsutane put it, 'Japanese differ completely from, and are superior to, the peoples of China, India, Russia, Holland, Siam, Cambodia, and all other countries of the world, and for us to have called our country the Land of the Gods was not mere vainglory.'[23] The Great Wall of China represented, in part, a comparable judgement, as did the prohibition in India on orthodox Brahmins travelling overseas. For Europeans as they began to explore the world, it was natural, at least for some, to take Aristotle as guide in regarding those they met as miniature men, homunculi, who were therefore as inferior to the Spaniards 'as children are to adults and as women are to men', being 'barbarians lacking in reason, and in knowledge of God, and in communications with other nations'.[24] The very word 'barbarian', which had originally meant 'non-Greek-speaking', came to mean, after the Persian War, brutal, uncivilized and 'hateful to the gods'.[25]

Less extreme has been the recognition that other cultures may be inferior overall, but that they still have something, or even much, to teach others. The same Hirata Atsutane who affirmed the superiority of the Japanese over the Dutch, nevertheless recognized the importance of Dutch scientific research (in method as much as in content), tied as it was to trade and exploration:

In Holland, one of the countries of Europe (though a small one), they consider astronomy and geography to be the most important subjects of study because unless a ship's captain is well-versed in these sciences it is impossible for him to sail as he chooses to all parts of the world. Moreover, the Dutch have the excellent national characteristic of investigating matters with great patience until they can get to the very bottom.[26]

It is thus possible for people to maintain the superiority of their own culture and yet to recognize virtue in that of others. This increasingly happened in nineteenth-century Europe and America, when people came to recognize that Asia had much to teach them: *ex oriente lux*. Tennyson, in his poem 'Akbar's Dream', admired the attempt of Akbar[27] to establish what in Tennyson's view nineteenth-century England much required, a 'faith of Reason' integrating different and conflicting religions:

> I can but lift the torch
> Of Reason in the dusty cave of Life
> And gaze on this great miracle, the World:
> Adoring That who made, and makes, and is,
> And is not, what I gaze on – all else, Form,
> Ritual, varying with the tribes of men.[28]

But for all that India might teach the West by way of wiser religion, Tennyson had no doubt that the salvation of India lay with his own superior culture:

> From out the sunset poured an alien race,
> Who fitted stone to stone again, and Truth,
> Peace, Love and Justice came and dwelt therein,
> Nor in the field were seen or heard,
> Fires of Suttee, nor wail of baby-wife,
> Or Indian widow.[29]

At the other extreme, beyond the recognition that other cultures may have wisdom to impart (even though, in general, they remain defective), lies the recognition that other cultures may be superior to one's own. That certainly affected Asia at the beginning of the twentieth century, when Hirata Atsutane's admiration of Dutch science became far more general, and carried with it an attack by Asians on their own cultures because they were blocking the development of science. In China, Chen Du-xiu (1879–1942) and Wu Zhi-hui (1865–1953) were both leading advocates of materialism, and this led them to make what Kwok has called 'frenzied attacks on tradition'.[30]

> This odorous thing, the national heritage, thrives along with concubinage and opium. Concubinage and opium in turn thrive together with status – and wealth – consciousness. When national learning was at a height, all politics were rotten. This is because Confucius, Mencius, Lao-tzu, and Mo-tzu were

products of the chaotic world of the Spring and Autumn and the Warring
States. They must be thrown into the latrine for thirty years. ... What is national heritage and what bearing has it on our present world? It is nothing
more than a relic of the world, worthy of preservation and nothing more.[31]

Cultural relativity differs from all three of these ways of handling cultural
difference. Even more, it challenges all three, because it argues that cultures are incommensurable. They cannot legitimately be compared with
each other because the standards of measurement are not absolute. They
come necessarily from one culture or another. Benjamin Franklin could
thus take a very different view of the North American Indians:

> Savages, we call them, because their manners differ from ours, which we think
> the perfection of civility. They think the same of theirs. Perhaps, if we could
> examine the manners of different nations with impartiality, we should find no
> people so rude as to be without rules of politeness; nor any so polite as not to
> have some remains of rudeness.[32]

If, then, standards of judgement are culturally relative, the hope of finding
independent courts of appeal seems remote. What seems obvious on one
side of the Pyrenees will not seem so on the other, as Pascal put it, anticipating the exploration of cultural relativity in the eighteenth century in
works that put the critique of European societies in the words of observers
from 'the other side of the Pyrenees'. Of these, Montesquieu's Les Lettres
Persanes (in which a supposed Persian writes letters home about the extraordinary nature of Parisian society) was the best known, but there were
others who made use of cultural relativity to criticize their own society.
The Chinese protagonist in Oliver Goldsmith's The Citizen of the World,
which appeared in 1762, makes exactly this point:

> When I had just quitted my native country, and crossed the Chinese wall, I
> fancied every deviation from the customs and manners of China was a depart-
> ing from nature; I smiled at the blue lips and red foreheads of the Tonguese;
> and could hardly contain when I saw the Daures dress their heads with horns;
> the Ostiacs powdered with red earth; and the Calmuck beauties, tricked out
> in all the finery or sheep-skin, appeared highly ridiculous; but I soon perceived
> that the ridicule lay not in them, but in me; that I falsely condemned others
> for absurdity, because they happened to differ from a standard originally
> founded in prejudice or partiality.[33]

Already in the eighteenth century ideas of relativity were making the
ground tremble beneath the feet of those who liked to be certain where
they stood. As Thomas Reid complained about David Hume's system, it
'leaves no ground to believe any one thing rather than its contrary'.[34] The

savage becomes once more 'the noble savage' of the Golden Age, living in an unfallen paradise of natural innocence.

In a general way, therefore, ideas of relativity had been familiar for a long time, and were already raising questions about the chance of finding an independent court of appeal in history. Those general ideas received dramatic and powerful reinforcement in that same year, 1917, in which Henson found himself in trouble. That was the year in which Einstein published 'Cosmological Considerations on the General Theory of Relativity', the famous paper in which he added the cosmological constant to his original equations when he realized that they were not yielding a solution representing a static universe. Stephen Hawking, putting 'the universe in a nutshell', summarized the issue in this way:

> Einstein's general theory of relativity transformed space and time from a passive background in which events take place to active participants in the dynamics of the universe. This led to a great problem that remains at the forefront of physics in the twenty-first century. The universe is full of matter, and matter warps spacetime in such a way that bodies fall together. Einstein found that his equations didn't have a solution that described a static universe, unchanging in time. Rather than give up such an everlasting universe, which he and most other people believed in, he fudged the equations by adding a term called the cosmological constant, which warped spacetime in the opposite sense, so that bodies move apart. The repulsive effect of the cosmological constant could balance the attractive effect of the matter, thus allowing a static solution for the universe. This was one of the great missed opportunities of theoretical physics. If Einstein had stuck with his original equations, he could have predicted that the universe must be either expanding or contracting. As it was, the possibility of a time-dependent universe wasn't taken seriously until observations in the 1920s by the 100-inch telescope on Mount Wilson.[35]

When that happened, and after Hubble had demonstrated the expansion of the universe, Einstein made the remarkable admission to George Gamow, that 'the introduction of the cosmological term was the biggest blunder he ever made in his life'.[36] Ironically, it seems now that it was not the blunder that he imagined, because he may have been using a premature language to speak of the effect of vacuum energy which acts just like the cosmological constant. Even so, this episode illustrates several of the themes already raised in this book: the corrigibility of science; the power of existing 'grand narratives' to lead to what Hawking called 'a fudge', and the need, therefore, to produce warrants for one's assertions.[37]

However, in 1917 that was not the issue. What Einstein's paper did was to accelerate the ways in which the theory of general relativity could be applied to cosmological questions; and one effect of that was to take relativity, not as a scientific theory, but *as a metaphor*, very extensively into educated imagination. That is what had such devastating consequences for the hope that history might serve as an independent court of appeal in conflicts of interpretation, because it introduced relativity into the under-standing of both time and history *as a metaphor*.

Extremely few people at the time understood relativity[38] beyond a rather vague impression that where the dimension of time is combined with the three dimensions of space to form space-time, time and space do not exist independently of the universe or of each other, and that this creates in consequence the many puzzles of time – that someone who goes off into space comes back younger than those who stayed on earth; that since individuals pursue their own trajectories through space-time, only to two people at rest with respect to each other will their times agree; that of two identical clocks in two planes circling the earth, the one in the plane flying east has recorded slightly less time; and so on.

Relativity, however, as a metaphor became so powerful that it has ended up as a domain assumption of the postmodern world.[39] Even before that, as early as 'Little Gidding' in 1942, T.S. Eliot was making use of the metaphor, without much understanding of what was going on:

> Here, the intersection of the timeless moment
> Is England and nowhere. Never and always. ...
> The moment of the rose and the moment of the yew tree
> Are of equal duration. A people without history
> Is not redeemed from time, for history is a pattern
> Of timeless moments.[40]

It sounds impressive. But what does it actually mean? In a letter to *The Times*, Sir George Engle recalled:

> I was present when in 1943 Eliot gave a talk on poetry to the Sixth Form at Charterhouse and at the end was asked by a 'history specialist' (J.V. Judah) what he meant by those words. After a long pause, Eliot replied: 'I have really no idea.'[41]

In more exact terms, the philosophers of time were not doing much better. There is a story told of one of them, the renowned G.J. Whitrow. He was stopped in London by a Japanese tourist, whose command of English was not very good, and who asked him, 'Please, what is time?', and Whitrow replied, 'That's a very difficult question, and I'm not sure I can answer it.'

It is, therefore, not surprising that few lay people understood or were even much involved in Einstein's 'difficult questions'. So it was not the general theory but rather the metaphor of relativity that has had such a destructive effect on the appeal to history. It has made popular a sense that both culture and time are relative to individual observers.

Where culture is concerned, it produced various theories of cultural relativity. Cultural relativity holds that cultures are tied to particular societies, each of which has its own history, geography and circumstances. To a great extent, each society with its culture pursues a history of its own, independent of the particular individuals who happen to live at any one time. In the characteristically succinct observation of Tacitus, 'Principes mortales, rempublicam aeternam esse.'[42] Of course some individuals are necessary, but their own beliefs and practices are formed by the culture they inherit at least as much as they in turn form it. Some have argued more strongly that culture actually determines what people believe and do, because cultures are necessarily well adapted to the circumstances in which they exist, whether these happen to be the asphalt or the Amazon jungle. If they were not well adapted, they would have changed or gone to extinction.

It is then also argued that there is no independent ground on which to stand in order to judge whether one culture is superior to another. The best a historian or an anthropologist can do is to try to make value-free descriptions of the past or of the present, as the anthropologist R. Lowie argued in 1937, in his book *History of Ethnological Theory*.[43] Lowie accepted Henson's point that no one can escape 'natural and unavoidable bias and tendency',[44] so that 'the anthropologist as an individual cannot but respond to alien manifestations in accordance with his national and individual norms ...'.[45] The aim, however, must be to make a neutral and value-free description and explanation of what has been observed: '... as a scientist, however, he merely registers cannibalism or infanticide, understands, and if possible explains such customs'.[46]

That may seem an admirably tolerant attitude to adopt, but cultural relativism actually questions its possibility. Yes, it accepts that cultures cannot be measured against each other in order to determine which is best, because the questions always have to be asked, 'Best for what?' and 'Best according to whom?' But the sharp edge of relativity is that we can never escape the consequence of the culture in which we happen to live, either in choosing what to study, or in deciding how to make that study. Our descriptions are always 'value-laden', because we can never entirely escape from the values of the culture in which we have been brought up and live. As Shweder and Bourne point out:

> The relationship between what one thinks about (e.g., other people) and how one thinks (e.g., 'contexts and cases') may be *mediated* by the world premise to which one is committed and by the metaphors by which one lives.[47]

It follows that if the controlling metaphor is one of relativity, then there cannot be an appeal to 'history' as though it exists objectively and escapes the relativity of the accounts that historians give of it.

Equally subversive was the way in which the metaphor of relativity emphasized that time is relative to observers, and that consequently there is no fixed quantity that pours forth from the Creator of space and time, in which that Creator writes history as a providential story. If *that* had been so, then the work of historians would be simple: it would be to trace and to retell that providential story – exactly indeed as a Fellow of my first Cambridge College, Corpus Christi, did in 1839, in two volumes entitled *The Theocratic Philosophy of History; Being an Attempt to Impress upon History Its True Genius and Real Character; and to Present it, Not as a Disjointed Series of Facts, but as One Grand Whole, Exhibiting the Progress of the Social System, Tending Under the Conduct of Its Divine Author, by Gradual and Almost Imperceptible Advances to its Completion.*[48] The title may be long, and in his preface Schomberg, the author, apologized for it being so. The point, however, is not the length of the title, but the fact that that attempt can no longer be made. Even if one believes that history moves to some point, or *telos* (that it is, in other words, teleological), it is still entirely possible to keep God out of it. That is exactly what the anthropologist Alfred Lewis Kroeber argued in 1915: 'The causality of history is teleological. ... Teleology of course does not suggest theology to those free from theology.'[49]

Anthropology and the Appeal to History

That remark of Kroeber's is important because it is a reminder that arguments about history as a court of appeal were not confined to Henson and his opponents. The debate about history played a major part in the early development of Anthropology as an academic discipline. Kroeber is often regarded as one of the great 'founding fathers' of Anthropology, at least of the second generation, and that remark about teleology was one of his famous 'eighteen professions'. The 'eighteen professions' were published by him in 1915 to lay down the fundamental ground-rules of Anthropology – and to lay them down, therefore, very much under the influence of Franz Boas, because Kroeber was one of his earliest research students.

Boas certainly was one of the founding fathers of Anthropology. In almost all respects Kroeber's 18 professions reproduce the way in which Boas had set out to establish Anthropology – except for one: the sixth profession reads, 'The personal or individual has no historical value, save as illustration' – the exact opposite of what Henson (following Carlyle, p 13) believed.

In itself, the sixth profession seems an innocuous remark, but yet again it marks the way in which Anthropology during Henson's lifetime was already subverting his appeal to history, because it was raising even more relativizing questions about the nature of time and the meaning of history. For it was in that same stormy year of 1917 that Kroeber published his *California Kinship Systems*[50] – an opening shot in what became a running battle with Boas about the status and worth of history, and about whether there can or should be 'an appeal to history as an integral part of Anthropology'.

The importance of this in relation to Henson's project of making history an integral part of Christian apologetic is obvious: Boas faced exactly the same problem of objectivity as Henson did, and at much the same time, although of course the particular issues were different.

Boas started work at a time when the embryonic discipline of Anthropology was dominated by what postmodernists would call meta-narratives, but which early anthropologists were more inclined to call 'grand theories'. These were overriding theories which were then illustrated by examples and anecdotes, the so-called 'just-so' stories which cannot possibly be refuted because they admit only the evidence that supports their case. An early example of these (though one that is still with us) was the 'just-so' story that uses evolution as the controlling narrative, for which confirming evidence is inevitably found. It remains true to the present day that anything *can* be given an evolutionary explanation in terms of natural selection and survival, not least because anything that *has* survived must have a reason for its survival.

Transferred to the emerging discipline of Anthropology, the 'just-so' story of evolution produced proposals every bit as absurd as the recent proposals about memes.[51] For example, the first president of the American Anthropological Society, W.J. McGee, attempted in 1895 to account for the success of the Anglo-Saxons by citing the cultural advantage of the English language:

> Possibly the Anglo-Saxon blood is more potent than that of other races; but it is to be remembered that the Anglo-Saxon language is the simplest, the most perfectly and simply symbolic, that the world has ever seen, and that by means of it the Anglo-Saxon saves his vitality for conquest instead of wasting it under the Juggernaut of a cumbrous mechanism for conveyance of thought.[52]

Boas fought against grand theory of that kind – the equivalent of Henson facing Gore. Boas resisted large theories imposed without appeal to fact, without regard to evidence – a position actually celebrated in a book review in *American Anthropologist* in 1904:

> [The book] is purely theoretical, and no facts whatever are adduced. This is not a criticism of the book. In fact it is one of the beauties of it. The book is not large enough to record the facts, and its style and character would have been changed, I had almost said spoiled, by their introduction. Any one who reads the book can see that the author's head was full of facts and that all he was trying to do was to reason from a store of facts to certain large conclusions. Those who speak disrespectfully of this method are often unable to make any use of their facts, however many they possess.[53]

Boas went in exactly the opposite direction: trained in physics, he brought to Anthropology a hard-headed, Baconian commitment to induction, to the gathering of evidence and facts before the elaboration of theory – exactly what Henson wanted in the case of history. In 1902, Boas wrote of what he called 'the general scientific principle'

> ... that the sound progress of science requires of us to be clear at every moment, what elements in the system of science are hypothetical and what are the limits of that knowledge which is obtained by exact observation. ... There are but few students who possess that cold enthusiasm for truth that enables them to be always clearly conscious of the sharp line between attractive theory and the observation that has been secured by hard and earnest work.[54]

'Hard and earnest work': exactly those qualities that had made Henson turn down the attractive offer of the Regius chair (p 13). What Henson called 'the patient and infinitely laborious historian' had an exact counterpart in Franz Boas. Both of them had such a profound suspicion of grand theory that they fall easily into Lyotard's definition of postmodernism as 'incredulity towards grand narrative' – what Boas called 'attractive theory'.

It may seem odd to regard Henson and Boas as postmodernists, partly because they were themselves embedded in narratives on a very grand scale indeed, Henson in Christianity and Boas in post-Baconian science, partly also because Boas died in 1942 and Henson in 1947. But Lyotard of course claimed that the 'post' in postmodernity is not a matter of chronology but of attitude, an attitude that calls in question the acceptance of *any* grand narrative as though it is secure against radical question or amendment. For both Henson and Boas, an amendment to the grand narratives of their time was demanded by a far more serious attention to history.

But what kind of history? That may seem a fairly trivial question, but in truth it is of paramount importance. It split Henson from Gore. It also split Boas from Kroeber, in a way that has had an enduring consequence, not just for the emergence of Anthropology, but equally for the emergence of Sociology, and for its internal debate about whether it is or should aspire to be a science. Should the letters SSRC stand for 'the Social Science Research Council', or, as Keith Joseph insisted when he was education minister, 'the Social *Studies* Research Council', on the grounds that Sociology and Anthropology are arts and not sciences? Throughout the twentieth century, battle was joined in the social sciences – or social studies – between positivist and hermeneutic accounts. Positivistic anthropologists and sociologists sought to bring the reasons why things happen under general or covering laws; hermeneutic theorists argued that humans behave in ways that will never be brought completely under general law descriptions. As Richard Miller has pointed out, the side taken in this battle determined the kind of work that anthropologists or sociologists actually did.[55]

The importance of the issue for Boas can be seen in the controversy about diffusion: where similar cultural items are found in geographically separated societies (such things as shamanism or burial practices implying belief in a life of some sort beyond death), is it the case that they have diffused from a single original discovery, or have they been discovered independently in several different places? Boas did not deny that there are many recurrent items of culture that cannot be explained by diffusion (that seem, in other words, to have arisen in disparate societies without connection to each other). So how do we account for these items that arise independently of each other, and yet resemble each other? Not, according to Boas, by appeal to some grand and overarching process, some causative principle like evolution or, in the case of Gore and the Christian Church, like providence and episcopacy. This kind of 'grand narrative' argument Boas summarized thus in the case of Anthropology:

> The fact that many fundamental features of culture are universal, or at least occur in many isolated places, interpreted by the assumption that the same features must always have developed from the same causes, leads to the conclusion that there is one grand system according to which mankind has developed everywhere; that all the occurring variations are no more than minor details in this grand uniform evolution. It is clear that this theory has for its logical basis the assumption that the same phenomena are always due to the same causes.[56]

In contrast, Boas insisted that a proper attention to detail, or to the facts in each case, demonstrates that histories are indeed idiosyncratic, and not the product of some preceding and universal cause:

Therefore we must also consider all the ingenious attempts at constructions of a grand system of the evolution of society as of very doubtful value, unless at the same time proof is given that the same phenomena must always have had the same origin. Until this is done, the presumption is always in favor of a variety of courses which historical growth may have taken.[57]

On that basis, Boas certainly regarded the evidence of history as central to an anthropologist's task. But, to repeat the point: what kind of history? As early as 1887, Boas had distanced himself from those historians who, as he put it, 'dwell admiringly on the character of their heroes, and take the most lively interest in the persons and nations they treat of, but are unwilling to consider them as subject to stringent laws'.[58]

'Laws': that nomothetic ambition which possessed so many at the end of the nineteenth century, offering the hope that history would turn into a science, as the economist Alfred Marshall summarized the point in 1897:

Social science, or the reasoned history of man, for the two things are the same, is working its way towards a fundamental unity; just as is being done by physical science, or which is the same thing, by the reasoned history of natural phenomena.[59]

Thus for Boas, if Anthropology was to become a *science*, it should prove possible to discern, not one single cause, but laws operating within the histories available for study. Way back in 1888, Boas had stood on a peak in Baffin Land and had glimpsed a new world:

The frequent occurrence of similar phenomena in cultural areas that have no historical contact suggests that important results may be derived from their study, for it shows that the human mind develops everywhere according to the same laws. The discovery of these is the greatest aim of our science. To attain it many methods of inquiry and the assistance of many other sciences will be needed. Up to this time the number of investigations is small, but the foundations have been laid by the labors of men like Tylor, Bastian, Morgan and Bachofen.[60]

So Boas reached the ambition of what he called 'my life's task':

Thus arose my plan to regard as my life's task the [following] investigation: How far may we consider the phenomena of organic life, and especially those of the psychic life, from a mechanistic point of view, and what conclusions can be drawn from such a consideration?[61]

Boas put the hope for identifiable laws like this:

When we have cleared up the history of a single culture and understand the effects of environment and the psychological conditions that are reflected in it we have made a step forward, as we can then investigate how far the same causes or other causes were at work in the development of other cultures. Thus by comparing histories of growth general laws may be found. This method is much safer than the comparative method, as it is usually practised, because instead of a hypothesis on the mode of development actual history forms the basis of our deductions.[62]

There could hardly be a stronger statement of the appeal to history as an integral part of scientific Anthropology. Yet this is the man against whom Kroeber, his star pupil, turned and accused of not being a historian at all. All that Boas had done, claimed Kroeber in 1935, was to use an appeal to history as a weapon against the tellers of those just-so stories:

> When he came on the scene, he found anthropology taken up with schematic interpretations – Morgan's will serve as a typical example; and he unhesitatingly proceeded to show that these schemes seemed valid only as long as the fact was ignored that they were built up of subjectively selected pieces of evidence torn out of their historical context, that is, their actual context in the world of nature. In his insistence that this context may not be violated, Boas may have seemed, possibly even to himself, to be following historical method. But it was merely historical method applied as a critical safeguard; the problems with which he concerned himself were not historical except in minor cases, but concerned with process as such.[63]

Kroeber in contrast regarded *himself* as the historian, arguing fiercely and publicly that a true historian would never subordinate his work to science: Boas does not *do* history, he wrote, putting the word 'do' in italics. In Kroeber's view, 'the historic approach, as distinct from mere historical technique', does not deal with the dimension of time alone but combines it with the dimension of space, integrating the two into a single space-time dimension[64] – a good example, incidentally, of Einsteinian phrases drifting loosely into other kinds of discourse. He summarized his argument in this way:

> The essential quality of the historical approach as a method of science I see as its integration of phenomena into an ever widening phenomenal context, with as much preservation as possible – instead of analytic resolution – of the qualitative organization of the phenomena dealt with. The context includes the placing in space and time and therefore, when knowledge allows, in sequence. But I see narrative as incidental rather than as essential to the method of history in the wider sense. Recognition of quality and of organizing pattern seems much more important. This is unorthodox but appears to me to be cardinal.[65]

It led Kroeber to conclude, in this same year of 1917, that history is determined by cultural patterns and not by individuals. Oh yes? Then what about the three individuals who, quite separately, rediscovered

Mendel's principles of genetics, three people called DeVries, Corens and Tschermak? Many regarded this as chance, as being, in Kroeber's words, 'a meaningless play of capricious fortuitousness'. Kroeber in contrast argued that an event of this kind revealed, as surely as the rise and fall of hem-lines in women's fashion, what he called

> a great and inspiring inevitability which rises as far above the accidents of per-sonality as the march of the heavens transcends the wavering contacts of random footprints on clods of earth. Wipe out the perception of DeVries, Correns, and Tschermak, and it is yet certain that before another year had rolled around, the principles of Mendelian heredity would have been pro-claimed to an according world, and by six rather than three discerning minds.[66]

What Kroeber meant by the title he gave to that article, 'the *super*-organic', was the transcendence of individual organisms by the circumstances in which they live. And that is why he added that sixth item to his profes-sions of anthropological faith: 'The personal or individual has no historical value, save as illustration.'

However, within the year, that same stormy year of 1917, Edward Sapir had already replied with the question, 'Do We Need a "Superorganic"?':

> Shrewdly enough, Dr. Kroeber chooses his examples from the realm of inven-tions and scientific theories. Here it is relatively easy to justify a sweeping social determinism in view of a certain general inevitability in the course of the acquirement of knowledge. This inevitability, however, does not alto-gether reside, as Dr. Kroeber seems to imply, in a social 'force' but, to a very large extent, in the fixity, conceptually speaking, of the objective world. This fixity forms the sharpest of predetermined grooves for the unfolding of man's knowledge. Had he occupied himself more with the religious, philosophic, aesthetic, and crudely volitional activities and tendencies of man, I believe that Dr. Kroeber's case for the non-cultural significance of the individual would have been a far more difficult one to make. ... One has only to think seriously of what such personalities as Aristotle, Jesus, Mahomet, Shakespeare, Goethe, Beethoven mean in the history of culture to hesitate to commit one-self to a completely non-individualistic interpretation of history.[67]

It seems an obvious criticism. It is nevertheless ironic that this criticism came from the very man who was to become one half of the famous Sapir–Whorf hypothesis. It is ironic, because in one of its forms, that hypothesis argues that the nature of a language dictates the thought, perception and behaviour of its speakers. It is a linguistic instead of a social determinism, in which, as Sapir put it in 1929, people are *at the mercy* (strong phrase indeed, 'at the mercy') of language:

Human beings do not live in the objective world alone, nor alone in the world of social activity as ordinarily understood, but are very much at the mercy of the particular language which has become the medium of expression for their society. It is quite an illusion to imagine that one adjusts to reality essentially without the use of language and that language is merely an incidental means of solving problems of communication or reflection. The fact of the matter is that the 'real world' is to a large extent unconsciously built up on the language habits of the group.[68]

On that basis, we arrive inevitably at linguistic relativity, because, to quote him again, 'the worlds in which different societies live are distinct worlds, not merely the same world with different labels attached'.[69] So the nature of time and the meaning of history are differently understood in different societies, and anthropologists were soon producing examples. Whorf himself claimed that the Hopi in North America cannot have the same sense of time as ourselves because their language does not allow it:

After long and careful study and analysis, the Hopi language is seen to contain no words, grammatical forms, constructions or expressions that refer directly to what we call 'time', or to past, present, or future, or to enduring or lasting, or to motion as kinematic rather than dynamic. Hence, the Hopi language contains no reference to 'time', either explicit or implicit.[70]

This does not mean that the Hopi have no sense of relationships in space and time. It means simply that they live in a different metaphysical construction of those relationships, and therefore they live a different experience of what we would call 'time' from our own perspective and construction:

The Hopi language and culture conceals a *metaphysics*, such as our so-called naïve view of space and time does, or as the relativity theory does; yet it is a different metaphysics from either.[71]

If in consequence 'the Hopi language gets along perfectly without tenses for its verbs',[72] and if therefore the Hopi live happily without any sense of history like our own, so do many others. U Wun claimed exactly this for the Burmese:

Burmese is a tenseless language. Past, present and future are not so important, the language talks about actions regardless of the time they take place in. Past and future find their reality only in the present. ... That is why most Burmese are not happy if they have to make scheduled appointments. For them time is relative and subjective, the time of cooling a pot of rice, or the time the sun is about to set, changing with season and human intention.[73]

Note the words, 'For them time is relative and subjective'. So it is for all human communities in relation to their languages. Many of them have no word for history, and therefore no idea to what it is that Henson or Boas or Kroeber want us to appeal. It was this that led Lévi-Strauss to suggest, in the last chapter of *The Savage Mind*, that we need a different contrast:

> The clumsy distinction between 'people without a history' and others could with advantages be replaced by a distinction between what I called 'cold' and 'hot' societies: the former seeking, by the institutions they give themselves, to annul the possible effects of historical factors on their equilibrium and continuity in a quasi-automatic fashion; the latter resolutely internalising the historical process and making it the moving power of their development.[74]

If that is so, then the effect is obvious and dramatic: it relativizes history in a radical way, because it relativizes the human understanding of time. If it is true, as Lévi-Strauss argued, that members of what he called 'cold' societies do not accept the past as past, they cannot be concerned with whatever or whoever governs the change from a prior state to another, later in time. Instead, as he put it, they try, 'with a dexterity we underestimate, to make the states of their development which they consider "prior" as permanent as possible'.[75]

Whose History?

This unfolding, though now distant, history of early Anthropology is simply one example of the way in which Henson's programme of appealing to history for adjudication on contested issues, and in order to challenge 'spin' and propaganda, came under threat, even during his lifetime. The threat would be even greater if we looked at that humming hive of historians who have been writing on the meaning of history during the last half century – of whom there have been many. Even if one allows the objectivity of evidence, the facts to which Henson wished to appeal, the interpretation of evidence, is a matter of interests. Even Walsh's 'colligation' which began as a matter of tracing intrinsic relations between events and locating them in their historical context[76] soon became the grouping of events under appropriate dominant concepts and leading ideas; true, those concepts and ideas must not be chosen arbitrarily but must arise from the facts and must make the past 'real and intelligible to us', to quote from his essay on 'Colligatory Concepts in History'; but in fact this is already a concession to subjective choice, since, as Hayden White observed, 'There are no grounds to be found in the historical record itself for preferring one way of construing its meaning over another.'[77]

Given the relativity of human understandings of history and of what history is, the question then becomes inevitable: to whose history and to what history might an appeal be made? If the appeal to history is to destroy error and defeat one's enemies, or if it is to summon up defence against, not just a bishop of Oxford, but against the far more sinister proponents of fascism such as Mussolini, then the history needed will surely be – will it not? – that of Benedetto Croce, whose *Teoria e storia della storiografia* (*Theory and History of Historiography*) was also published in that same stormy year of 1917. For it was his understanding of the meaning of history that led him inexorably into an opposition to Mussolini. It was an opposition at least as emphatic as that of Henson in the House of Lords, but at far greater risk to himself. When the Lateran Pact or Treaty was made in 1929 between the Vatican and the Fascist government, Croce found himself, like Henson on that other occasion (p 15), in a minority of one: his was the only public speech on record to denounce the Pact.

But was *his* the history to which Henson would have wished to appeal? It seems doubtful. Croce distinguished sharply between 'chronicle' and 'history': 'chronicle' is a setting in serial sequence of such remains as there are from the past, more or less what Henson meant by 'fact'. But Croce did not think too highly of that. In writing *history*, Croce argued, historians will enter into the past and relate it to the present by rethinking it: they will, as he preferred to put it, make it contemporary. In a famous passage in his *Theory and History*, he asks:

> Do you wish to understand the true history of a Ligurian or a Sicilian neolithic man? First of all, try if it be possible to make yourself mentally into a Ligurian or Sicilian neolithic man; and if it be not possible, or you do not care to do this, content yourself with describing and classifying and arranging in a series the skulls, the utensils, and the inscriptions belonging to those neolithic peoples. Do you wish to understand the history of a blade of grass? First and foremost, try to make yourself into a blade of grass, and if you do not succeed, content yourself with analysing the parts. ... This leads to the idea from which I started in making these observations about historiography, as to history being *contemporary* history and chronicle being *past* history.[78]

The thought of Henson sitting in his book-lined study at Durham turning himself into a Ligurian neolithic man, let alone into a blade of grass, defies the imagination. One *can* see him, maybe, classifying and arranging – that close and laborious attention to evidence which he regarded as the mark of the new historian. But not, surely, for the purposes that Croce proposed, because Croce's understanding of history led directly to the grand theories of which Boas and Henson were so suspicious – evolution as the controlling metaphor of Anthropology, Gore's ecclesiology.

Croce, in contrast, was entirely clear that 'historical facts', those items to which Henson was inclined to appeal when challenged on the virgin birth and the resurrection, do not point to a foundation of permanent and independent truths on which one might build the case for Christianity. For Croce, there is no world except the human world, the world that humans construct. The belief that nature, or indeed the past, exists accessibly in independence from what we make of it ourselves is simply wrong – the belief that, as Lonergan said of the sixteenth-century Jesuits, Suarez, de Lugo and Bañez, 'they seem to have thought of truth as so objective as to get along without minds'.[79]

To Croce, in contrast, nature is simply a web of human concepts and categories serving human purposes. But that did not even remotely mean to Croce that among those purposes it is illegitimate to tell important and interesting stories, one of which is the story of (to pick up the name of the movement so detested by Henson) 'life and liberty'. That is why Croce countered the rise of Mussolini with his successive histories: in 1925, of the kingdom of Naples; in 1928, of Italy from 1871 to 1915; in 1929, of the baroque period in Italy; in 1932, of Europe in the nineteenth century, culminating, in 1938, in *History as the Story of Liberty*. If ever there was a grand theory controlling the facts, this surely was it – at least as grand as Gore's Anglo-Catholicism, and very far removed from what Henson regarded as 'an appeal to history'.

It leaves us, therefore, with the paramount question, whose history is the history to which in our conflicts we are supposed to appeal? The question has become all the more acute in our time. For we live, not only in the context of relativity in its applied and metaphorical senses (the subjectivity of ethical and of aesthetic judgements, for example), but also in the context of the deconstruction of privileged points of view.

As long ago as 1966, Jacques Derrida was speaking in words of which Croce would have approved, but Derrida was of course moving from them in a very different direction. At a conference in that year on 'The Languages of Criticism and the Sciences of Man', he accepted, like Croce, that we have nothing but the languages and concepts that we have inherited, even when we are at work knocking down the pretensions of the past. So he argued:

> *There is no sense* in doing without the concepts of metaphysics in order to attack metaphysics. We have no language – no syntax and no lexicon – which is alien to this history; we cannot utter a single destructive proposition which has not already slipped into the form, the logic, and the implicit postulations of precisely what it seeks to contest.[80]

Everything is, to use Croce's phrase, a web of human concepts and categories serving human purposes. We can only stand within the web in order to pull it apart – hardly what Henson saw as the historian's task.

On that occasion in 1966, Jean Hippolyte made the first response to Derrida's paper. He came to a final question in which he asked (by no means the last person to do so) whether he had understood what Derrida was talking about, or, as he put it, 'Were you getting at something else?'[81] In the course of his response, Hippolyte suggested that 'we have much to learn from the natural sciences', by which he meant that the norms of empirical verification were under threat: 'With Einstein, for example, we see the end of a kind of privilege of empiric evidence.'[82] Derrida emphatically agreed:

> The Einsteinian constant is not a constant, is not a center. It is the very concept of variability – it is, finally, the concept of the game. In other words, it is not the concept of some*thing* – of a center starting from which an observer could master the field – but the very concept of the game which, after all, I was trying to elaborate.[83]

It is startling to see how completely Einstein was misrepresented by both speakers. Both of them seem to have confused the word 'relativity' with the phrase 'all is relative', with the result that they have used the word and the concept of relativity in a way that is almost the exact opposite of what Einstein intended. In fact, the purpose of the theory of relativity is to 'formulate the laws of physics in a form in which they have absolute validity, independently of the state of motion of the observer'.[84]

Nevertheless, that response of Derrida is a neat summary of the deconstructive campaign against any philosophy, certainly any metaphysic, that claims a foundational point on which to stand and move the world. For Derrida immediately went on to make it clear that he did not mean by 'game' anything that might be legitimized by rules, unless at the same time it is remembered that the rules can be distanced from themselves by *différance*, as when William Webb Ellis picks up the ball and runs with it and invents the game of rugby; even more when croquet is played with flamingoes.

And here surely the project of finding independent courts of appeal in order to discern truth in the midst of conflict lies in ruins. For deconstruction does not simply denounce the pretension that there are some facts in the past to which we might gain access and which constitute 'history' behind or beyond the web of human interests. Deconstruction actually, and far more insidiously, subverts the notion of 'appeal' altogether, it destroys the very enterprise of appealing to innocent and

independent facts as a foundation for something else – a scientific Anthropology, for example, or a Christian apologetic.

So did Henson waste his money in setting up lectures to explore 'the
appeal to history'? Henson, it is true, had already foreseen the problem.
He knew perfectly well that human interests are paramount in everything
that humans undertake, including accounts of the past. What he hoped
was that a real distinction (a realistic distinction) might be made between
a non-negotiable core or essence of what is truly the case, and the interpretations placed upon it which are contingent and non-essential –
something like Franklin's 'rules of politeness and remains of rudeness' (p
20). There are some human experiences that are real to all who live, even
though the contingency of their occurrence is always different, exactly as
George Barker wrote of human suffering:

> I know only that the heart
>   Doubting every real thing else
> Does not doubt the voice that tells
>   Us that we suffer. The hard part
> At the dead centre of the soul
>   Is an age of frozen grief
> No vernal equinox of relief
>   Can mitigate, and no love console.[85]

Time certainly poses puzzles, but, as Henson once put it, at least the speed
of light does not betray us. Nor does the demand upon us of what he
called in his speech on Abyssinia, 'the demand of moral obligation' (p 13).

That is why he gave his Gifford Lectures in 1935[86] on 'Christian morality, natural, developing and final'. Here again the appeal is to history in
order to justify his claims:

> You will perceive that I am asking you to leave both the high latitudes of phil-
> osophy and the low lands of anthropology in which so many Gifford Lecturers
> have pursued their argument, and to confine yourselves to that limited scene
> in which, in spite of its limitation, all theories of faith and morals must finally
> come to judgement. In these Lectures we shall move strictly within the sphere
> of recorded history, for the materials of our argument can only be contributed
> by historic testimony, and the value of our conclusion can have no meaning
> outside historic conditions.[87]

It is also why he spoke, in his Fison Lecture on ethics and science, of what
he called the ultimate and sovereign moral law, 'the grand and cementing
principle of ordered society', which is developed and applied in different
ways in different societies. Without it, in Henson's view, societies will
have no ground on which to stand except the ground of science, because

science does at least offer warrants for its conclusions. And yet, as he looked at the moral philosophy of his time (see Chapter 4), he saw a complete retreat from anything like an appeal to moral law:

> The Moral Law itself, which is the grand cementing principle of ordered society, seems to be stricken with instability, and its claims, ceasing to be ultimate and sovereign, sink into a dependence on the conclusions of science.[88]

In a comparable way, he argued that there is a natural property of beauty: it can be expressed in varying ways in the many different forms of art, but the nature of beauty can be identified and described. It is, he claimed, what Ruskin had called in *Modern Painters*, 'the essence and the authority of the Beautiful and the True'.

Aesthetics and morality: here surely are exactly the two areas where, we are so often told, objectivity has been destroyed, and where subjectivity and relativism have seemed inevitable. People can appeal to Beauty and Goodness if they like, they can appeal to history, but they cannot expect everyone, or even anyone, to agree with them. Gore and Henson, Kroeber and Boas, can fight on forever, but they won't get a decision between them by an appeal to facts. In the postmodern world, the belief that there are independent facts to which one might make appeal in the case of disputes in history or aesthetics or ethics seems simply bizarre.

But I wonder. The recent explosion of work in the neurosciences and in neurophysiology throws much light on what happens in the human brain when it arrives at aesthetic and moral judgements. Taking that work into account, we can begin to see not only how we make the judgements that we do but also the reasons why they are related to facts independent of ourselves, if not in the way that Henson had hoped, at least in a way that belongs to the exposition of human truth. The next step, therefore, is to look at the appeal to beauty, and to make a start on that we will go with Henson into the dark corridors of Auckland Castle and stand beside him as he looks at the portraits of his predecessors.

3 THE APPEAL TO VALUE: ART IN CHINA AND THE WEST

In 1933, Henson was showing a visitor round Auckland Castle. To-gether they stood before the portraits of previous bishops of Durham, and Henson, in his usual pugnacious style, lamented their depressing characteristics – not the characteristics of his predecessors (though he surely had opinions about that), but those of their portraits as works of art. He called them brown soup in chipped bowls. That is a judgement about a work of art, an aesthetic judgement about the beauty or otherwise of certain objects (pictures hung on a wall) at which two people were looking.

Supposing his friend had disagreed and had said that the portraits were superb – caviar in gilded dishes: how could they decide who was right? They could both give reasons why they had reached their conflicting judgements, but reasons clearly could not arbitrate between them and bring them to a judgement with which they must both agree. Georgia O'Keefe (of whom Lillian Schacherl observed, in an aesthetic judgement, 'It would not be advisable to take Georgia O'Keefe's flowers to a tea party: they would dominate the conversation, reducing everyone to silence and diverting their attention to other – more – decadent thoughts'[1]) wrote in an exhibition catalogue in 1923:

> I found I could say things with color and shapes that I couldn't say in any other way – things that I had no words for. Some of the wise men say it is not painting, some of them say it is.

To what court of appeal could the wise men go for some decision between them? Henson had hoped that in making judgements in matters of his-torical truth, or of beauty, or of goodness, it should be possible to identify objective facts as distinct from subjective interpretations built upon them – historical facts, moral facts, aesthetic facts, independent of those who observe them.

It is a hope that seems now to lie in ruins. Among the demolition workers, the postmodernists have been prominent, but before them were the post-Humeans, since it was Hume who pointed out the fundamental objection to seeking independent and observer-free facts in these various areas. In the case of aesthetics he wrote:

> Euclid has fully explained all the qualities of the circle; but has not, in any proposition, said a word of its beauty. The reason is evident. The beauty is not a quality of the circle. It lies not in any part of the line whose parts are equally distant from a common centre. It is only the effect which that figure produces upon the mind, whose peculiar fabric or structure renders it susceptible of such sentiments. In vain would you look for it in the circle, or seek it, either by your senses or by mathematical reasonings, in all the properties of that figure.[2]

In the case of ethics the argument is the same: that we do not observe a property of goodness, but impose our evaluations as a matter of emotion and sentiment:

> If we can depend upon any principle which we learn from philosophy, this, I think, may be considered as certain and undoubted, that there is nothing, in itself, valuable or despicable, desirable or hateful, beautiful or deformed; but that these attributes arise from the particular constitution and fabric of human sentiment and affection.

With the consequence of this we have been living ever since: there are no properties of 'beauty' or 'goodness' that all people must see in the same way that they must see the property of circularity. Yes, certainly, we see things that stir our feelings, but only then do we begin to discuss among ourselves whether those things are or are not beautiful, are or are not good. With a circle we do not discuss it, we measure the radius.

Since, however, there are no measurements to be made that will demonstrate that a claimed value does indeed exist in some object, the conclusion seems inevitable that we cannot get values from facts, we cannot get an *ought* from an *is*.

One early defence was simply to claim that Hume was wrong and that the property of beauty actually does subsist (or fails to subsist) in works of art. But as Tilghman put it (in a book entitled very appropriately, *But Is It Art?*[3]), 'It is just not clear what could possibly count as a defining property of art'[4] – let alone of beauty:

> If it is not always possible to make sense of looking for likenesses or differences within a single art form, then the search for likenesses across the boundaries of art forms can appear distinctly bizarre. Try asking how Titian's *Venus of Urbino*, for example, is like 'Sweeney Among the Nightingales', 'Pavan for a Dead Princess', or the Taj Mahal, and that sense of the bizarre will surely register.[5]

So if there is no single property of beauty (or for that matter of goodness) that is independent of evaluation and is simply waiting 'out there' to be observed, then surely we have nothing other than those evaluations – those human judgements – to which we can appeal. We cannot appeal to beauty or to goodness, or for that matter to history, as though they are independent of our varying and conflicting evaluations.

So unhappy Henson! Yet again he finds himself in a minority of one, not this time in a contest against Gore, but in a contest against Hume, and his hopes of a rational and intelligent faith seem even more remote.

Henson, however, was not altogether wrong in either case. At the time when Hume lived, it had long been assumed that rationality and emotions were divided from each other. They were indeed at war with each other, in what Letwin has called 'the combat of reason and passion',[6] in which it is the task of reason to bring the emotions or passions under control.

Hume opposed this emphatically. He argued that the passions, whether calm or violent, were indeed independent of reason, but in such a way that reason cannot enter into them, still less control them. Passions arise from our interactions with the world around us, and they are, like heat and cold, simply what they are. Approval or disapproval of them or of their consequences is a sentiment that arises from the usefulness or otherwise of any action or character in contributing to the well being of society. It certainly does not arise from *observing* vice or virtue:

> When you pronounce any action or character to be vicious, you mean nothing but that from the particular constitution of your nature, you have a feeling or sentiment of blame from the contemplation of it. Vice and virtue, therefore, may be compared to sounds, colours, heat and cold, which, according to modern philosophy, are not qualities in objects but perceptions in the mind.[7]

Sentiment thus belongs to what Hume called 'the principle of humanity'. Reason participates in these appraisals: it is reason that rationalizes sentiment in the direction of judgement, and reason that adjudicates finally upon the vocabularies of beauty and of goodness. But the fact remains (for Hume) that reason cannot control the passions.

Hume thus preserved the distinction between reason and emotion, but in a way that was very different from what had previously been thought. Thus in searching for 'the general foundation of morals', the basic question for Hume was, to quote his own words,

> whether they be derived from *reason* or from *sentiment*; whether we attain the knowledge of them by a chain of argument and induction [reason], or by an immediate feeling and finer internal sense [emotion]; whether, like all sound judgement of truth and falsehood, they should be the same to every rational,

intelligent being [reason]; or whether, like the perception of beauty and de-
formity, they be founded entirely on the particular fabric and constitution of
the human species [emotion].[8]

It is this distinction between reason and emotion that leads inevitably to
the distinction between fact and value, between *is* and *ought*.

But what has changed dramatically in recent years is our understanding
of what Hume called in that passage 'the fabric and constitution of the
human species'. The change will not lead us back to finding 'beauty' or
'goodness' as independent properties and observable facts which everyone
who takes a look must see. However, it does lead in the direction of a
rational and shareable understanding of the reasons why we make the
judgements that we do in the area of ethics and aesthetics, and why the
vocabularies of beauty and goodness (and their deployment) are not
matters of subjective, or even private, opinion only.

Neuroscience, Emotion and Rationality

The background to this lies in the recent work in neuroscience and in
neurophysiology that ties together, in an entirely new way, perception,
emotion and rationality. It is this work that completely alters our understanding of 'the fabric and constitution of the human species', that is, our
understanding of ourselves and of the way we make historical, aesthetic
and moral judgements.

It is important to note from the outset that this new work accepts the
post-Humean conclusion that there is no *single* property of beauty or of
goodness lying within the objects of perception. However, it does not
accept that there are no independent properties open to observation
which lead to consistent judgements of value. Even more, this work shows
that emotion and rationality are *combined* in making these judgements of
value, so it raises a huge question mark against the whole post-Humean,
postmodernist, view that we simply add our own values on to initial
perceptions in a way that must be entirely relativistic and subjective,
because *inter*-subjective rationality does not enter into it.

In brief, therefore, the contrary argument will be:

- first, that our responses are often psychosomatic in an integrated way,
  and not always in a sequence of emotional perception followed by rational evaluation;
- second, that those integrated psychosomatic responses are tied to
  properties that do lie within the objects themselves, whether or not
  they are being perceived by some observer.

Let us begin then with aesthetic judgements. Here surely Hume and the postmodernists are right, that beauty lies in the eye of the beholder, not outside in the object. In contrast, however, consider the poet Norman Nicholson, standing one day in the Ruskin Museum in the Lake District, when two children from the village ran in. Having first, Nicholson recalled,

> knocked a vigorous, clashing, Bartok-like tune on the musical stones, they went over to a case of minerals, and pointed to a strange piece of quartz, golden, glittering and contorted as a Chinese lizard: 'That's my favourite,' said one. 'Smashing, isn't it?' replied the other. And from their direct and unperturbed response to natural beauty ... I felt that here, after all, were eyes that could see at least something of what Ruskin had seen.[9]

The conducive properties in that quartz – golden, glittering and contorted – evoked the language of satisfaction in the children, and not just in Ruskin alone. It does not follow that they have seen the object in the same way, or that they have seen the same object. As the critic Harold Osborne observed, 'Two men who look at the same picture may see two different pictures and two men reading the same poem may realize two different poems.'[10] Nevertheless, it is a direct seeing of an object that evokes the language of satisfaction. It is the reason why people are entirely right to look at something and to say exactly what they feel: 'Smashing, isn't it?' It is the profound justification for saying (despite often being told not to), 'I don't know much about art, but I do know what I like.'

'What I like' will be greatly changed and extended when I do know something about art. But that initial and immediate response, on which all art is parasitic, is important because it lies deeply embedded – or more accurately, deeply embodied – within us. How that is so is what we are now just beginning to understand.

But only *beginning*, because these are extremely early days in brain research. Immense progress has been made in mapping the geography of the brain and in studying the consequences of brain damage and deficit. Even so, our knowledge is still very limited, and we certainly do not know much in detail about the major human emotions – feelings, for example, of fear, pleasure, disgust, boredom, beauty, and indeed of God.

But we do at least know most about the emotion and feeling of fear. That is because it is easy to see when experimental animals are in a condition that humans would experience as fear, whereas it is virtually impossible to be sure that animals are in a condition that humans would experience as beauty.

The question is whether we can perhaps infer something about the emotion of beauty from what we know is happening when we experience the emotion of fear. If we can, this at once opens up an entirely different approach to understanding aesthetic judgements.

Imagine, then, that you are on holiday in India, and that as you walk along you see a snake lying on the path in front of you. Your immediate reaction is one of fear: you freeze on the spot, your heart beats a bit faster. But then – cautiously! – you take a closer look and you see that it is, after all, only a piece of rope – the classic example on which Indians rely in order to construct their theory of *maya*, of the imposition of our ideas on what we take to be the world outside.[11]

What has been going on inside your brain and body as you looked at the snake? External receptors in your eyes have transmitted sensory messages to specific areas of the thalamus, and these have then processed the signals and sent the results to specialized areas of the neocortex dealing with the higher stages of sensory processing, in this case of sight (of seeing the snake). The neocortex then sorts out what is going on and alerts the amygdala to initiate an appropriate response.

Among the many outputs from the amygdala are some long fibres descending along the ventral amygdalofugal pathways to the autonomic centres, and in this way cortically processed signals reach the brainstem so that our rapid and appropriate responses are initiated – freezing on the spot, heart beating faster. But since a rational scan of the situation is also involved, that fear of a snake may not even arise, because it is seen to be only a piece of rope.

That is a very over-simplified account, because it is isolating the amygdala in an artificial way, when we know that other brain regions, such as the orbitofrontal cortex, are involved. But at least it makes it clear that the amygdala does play a crucial part in emotional behaviour.

So what is the amygdala? In both halves of the brain, it is a small subcortical region in the anterior part of the temporal lobe, and it is called the amygdala because that is the Latin word for 'almond' and that roughly is its shape.

It has been known for a long time that bilateral damage to the amygdala leads to major disruption of emotional responses. The work of Weiskrantz[12] nearly 50 years ago examined monkeys in which the amygdala had been removed or damaged, and he found that the monkeys become very passive. They cannot distinguish between food and non-food items placed in front of them, and they put both into their mouths in order to examine them. In other words, by failing to associate visual and other stimuli with a primary reward or punishment ('primary' being one that has not been learned), they lack the appropriate feelings and emotions that food might otherwise evoke in a hungry animal. Similarly and more recently, the work of Damasio, LeDoux and many others has shown that lesions of the amygdala change or attenuate responses of fear.[13]

In humans, there is far less evidence, because we do not make direct experiments of that kind on each other. But where there has been damage or surgery to the amygdala in humans, the evidence does point in that same direction.[14] As early as 1939, lesions of the amygdala were being used to control violent behaviour, and it was observed in some patients that they expressed a strong sexual desire directed not necessarily towards a partner but to anything around them, including inanimate objects – a condition known in more general terms as the Klüver–Bucy syndrome, following their bilateral removal of the temporal lobe in rhesus monkeys.

More recently, R. Joseph recorded the plight of a young man whose amygdala had been removed in order to deal with severe seizures. As a result, he lost interest in people: although he could converse with them perfectly well, he was unable to interact with them emotionally.[15]

But in people where damage of that kind has not occurred, the central nucleus of the amygdala interacts with response control systems to produce appropriate responses. In the case of the snake, it might be freezing on the spot; in other circumstances it might be such things as jumping up and down with excitement, or weeping with grief.

This means that the amygdala plays an important part in processing the major responses to situations in which conducive properties are derived from the objects of perception – for example, the golden glittering contortion in that piece of quartz (p 42), or the snake on the path.

Conducive properties are so called because (from the underlying Latin, *duco*, I lead) they lead from the perceived object to one set of events within us rather than another, and they do this with a highly stable consistency: not only Ruskin but also Nicholson and the two boys saw them in the quartz, and so could we if we paid a visit to the Ruskin Museum.

So that 'source of stimulus' on the path in India carries within itself conducive properties that signal a danger, an object of fear, a snake; and those conducive properties lead via the amygdala to the consistent and appropriate emotion of fear: stress hormones are released, the heart rate changes, reflexes are initiated and a person may even be frozen on the spot.

In the case of fear, those events within us may actually *support* a Humean account of emotion being followed separately by rational evaluation, and that is so because it all happens so fast. That is exactly what Darwin found when he performed an experiment on himself in the London Zoo.

He wanted to find out whether he could use his reason to control his emotion of fear – mind over matter. So this is his account of what he did:

> I put my face close to the thick glass-plate in front of a puff-adder in the Zoo-logical Gardens, with the firm determination of not starting back if the snake struck at me; but, as soon as the blow was struck, my resolution went for noth-ing, and I jumped a yard or two backwards with astonishing rapidity. My will and reason were powerless against the imagination of a danger which I had never previously experienced.[16]

In the case of fear, Hume's account needs to be correct: if you stop to think about it, you may stop thinking, because the danger, if it is real, gets you.

In the process of evolution, therefore, the brain has been rewarded in terms of survival by retaining a direct pathway, a shortcut route, from the sensory thalamus into the amygdala, bypassing the neocortex, and initiat-ing output into the autonomic and motor nuclei – hence the speed of Darwin's reaction and the failure of his intended cortical control.

We have *retained* a direct pathway. The reason for the word 'retained' is that for most organisms the perception of conducive properties lying within the objects of perception is all they have. And therefore, not sur-prisingly, because survival is involved, the role of the amygdala is the same across species, even though the defensive behaviours that different animals adopt are different.

The implication, therefore, is obvious: that this brain system evolved very early on. It would certainly be rewarded in evolutionary terms: differ-ent species defend themselves against danger in many different ways, but the role of the amygdala is constant. Therefore, as LeDoux put it, 'When it comes to detecting and responding to danger, the brain just hasn't changed much. In some ways we are emotional lizards.'[17]

Maybe. But even then, the lizards that we happen to be, happen to be lizards in whom reason is not *always*, or even usually, bypassed in that shortcut, subcortical way. Even in the case of fear, the rational evaluation in the case of humans can be (and often is) part of the initial response. In any case, fear is not the only emotion, and Rolls in his book, *The Brain and Emotion*, criticized LeDoux for concentrating too exclusively on the one emotion, fear in rats, where the shortcut, subcortical route is indeed paramount. Rolls offered his own theory of the neural basis of emotion which he describes as 'conceptually similar' to that of LeDoux except in three respects: that LeDoux concentrates on the role of the amygdala in emotion and not on other brain regions such as the orbitofrontal cortex; that he concentrates on subcortical routes; and that he focuses mainly on fear-conditioning in rats.[18] In a later survey,[19] LeDoux accepted the criticism and conceded that, as he put it, 'essentially nothing is known about the neural basis of jealousy, envy, or trust, but at the psychological level these are believed to involve cognitive systems to a greater extent than basic emotions.'[20]

It is the involvement of cognitive systems that is so important. Despite that essential lack of knowledge to which he refers, it is already clear that in general, and far more often, there is a cooperation of emotions and reason in a single, coordinated response. Reason *and* emotion together review what is going on, so that human responses are formed from both, as we evaluate the signals from conducive properties in the objects of perception.

Thus emotions are now often defined as states produced by instrumental reinforcing stimuli, which may be by way of reward or punishment, as, for example, when people receive tokens of approval or disapproval, such as inclusion by way of being kissed, or exclusion by way of being spat at. Instrumental reinforcers are emotion-provoking stimuli which, because they vary and because they arise in contingently different circumstances, bring different emotions into being.

Some stimuli are not learned in the course of life, and are therefore known as primary (i.e. unlearned) reinforcers (for example, pain or the taste of food or of water). Others are not primary in themselves, but they become reinforcing because they become associated through learning with the primary reinforcers. This learning is known as stimulus reinforcement association, through which stimuli that have no intrinsic emotional effect become emotionally important – indeed, causative (or, to revert to the earlier language, pp 5–6, specifiable in the network of constraints).

From this, it is clear that emotions cannot be divorced from a cognitively accurate scan of the contingent context. Even at the level of the amygdala neurons involved, it is clear that neuron response can be modified by environmental considerations in some ways but not in others. So, for example, the work of Nishijo and others on four primate amygdala neurons has shown that response to a primary reinforcer (an item of food) diminished when salt was added to that item, so that the difference was cognized without the necessity of actual taste.[21] Wilson and Rolls have shown that there is a difference in primate neuron response between visual discrimination and rule-based behaviour, in the sense that the analysed neurons did not reflect the reinforcement value of visual stimuli when this was controlled by a rule, but they did respond to rewarding stimuli when there was a previous association between the visual stimulus and a primary reinforcer. This means that while the relevant amygdala neurons will respond to stimuli associated with primary reinforcers, they cannot be taught to do so by some other route – for example, by introducing an invariant rule of association, whereby entirely new stimuli always receive negative reinforcement, and familiar stimuli always receive positive reinforcement.[22]

Even this, however, does not mean, as used to be thought, that neural codes constitute a simple representation of the sensory world, since while it is true, as Fairhall and others put it, that 'the sequences of action potentials from single neurons provide an efficient representation of complex dynamic inputs, that adaptation to the distribution of inputs can occur in real time and that the form of the adaptation can serve to maximise information transmission', nevertheless the adaptive code is ambiguous, and this involves compromises:

> An adaptive code is inherently ambiguous: the meaning of a spike or a pattern of spikes depends on context, and resolution of this ambiguity requires that the system additionally encode information about the context itself. In a dynamic environment the context changes in time and there is a tradeoff between tracking rapid changes and optimizing the code for the current context.[23]

On a somatic scale, there has to be a comparable compromise between automatic responses to emotional stimuli and the rational evaluation of contexts, since otherwise we would be constantly hiding under the table or else sitting at it wondering why we are going to extinction. The point is that cognitive processing cannot be divorced from emotional experience, because it is involved in determining whether an event or stimulus arising in the external world is or is not reinforcing, in either of the two ways (primary or by association). The conducive properties must lie within the objects of perception, and not be a matter of imposing our own rules, since in that case the stable consistency of emotional experience could not occur as it does. The argument of LeDoux (Note 17), therefore, cannot, as a generalization, be right.

It is here that we begin to see the major difference in our understanding of what Hume called 'the fabric and constitution of the human species' (p 39). Hume might have been right about fear in some instances, but not about emotions in general, nor even about fear in *all* instances. The reason is simple: the signals from conducive properties do not usually reach the amygdala alone by the subcortical, bypass route. They arrive via the thalamus in the neocortex, and the appropriate areas in the neocortex review the situation. Even in the case of fear, they may come to the immediate conclusion that it is after all only a piece of rope, in which case the emotion of fear is not switched on.

On the other hand, if they *confirm* that the danger is real, the amygdala is instructed to initiate the appropriate emotional responses, not just in the brain but in the whole body. That is why a heart-stopping, stomach-churning, sweat-making experience in the presence of a snake may make people quake in their shoes. Emotions don't just happen in the head: the whole body gets involved because emotions have to do with our most important moments – not just with matters of life and death, but with all the profound satisfactions and revulsions that make us human. Thus it is *people* who experience emotions, not brains, however true it is that the brains humans have are the necessary condition for humans experiencing anything. As D.H. Lawrence (whose pictures ran into trouble with the magistrates, p 56) observed, 'You can't invent a design: you recognise it, in the fourth dimension – that is, with your blood and your bones, as well as with your eyes.'[24] There are many neuroscientific accounts of emotion (and of much else) which give the opposite impression, that brains are not just necessary but actually sufficient for the explanation of emotion, but those accounts are simply wrong.

Recognition and Response in Aesthetics and Art

And what has all this to do with the two children in the Ruskin Museum? Everything, because it is now obvious why they, and others, can look at something and immediately respond, 'That's smashing' – or 'That's beautiful', or whatever words they want to use: perhaps the word used by Bill Bryson when he went to see the Grand Canyon. When Bryson arrived, he couldn't see a thing: 'The fog was everywhere, threaded among the trees, adrift on the roadsides, rising steamily off the pavement: it was so thick I could kick holes in it.'[25] Still, when he saw a sign pointing to a look-out point about half a mile away, he decided to go there, 'mostly just to get some air':

> Eventually I came to a platform of rocks, marking the edge of the canyon. There was no fence to keep you back from the edge, so I shuffled cautiously over and looked down, but could see nothing but grey soup. A middle-aged couple came along and as we stood chatting about what a dispiriting experience this was, a miraculous thing happened. The fog parted. It just silently drew back, like a set of theatre curtains being opened, and suddenly we saw that we were on the edge of a sheer, giddying drop of at least a thousand feet. "Jesus!" we said and jumped back, and all along the canyon edge you could hear people saying "Jesus!", like a message being passed down a long line. And then for many moments all was silence, except for the tiny fretful shiftings of

the snow, because out there in front of us was the most awesome, most silencing sight that exists on earth.[26]

Conducive properties (derived from standing on the edge of the abyss) led
to the emotion and to the neural and bodily response of fear (jumping
back and shouting 'Jesus!'). That may be achieved by the shortcut, subcortical route straight into the amygdala; and if that is *all* that ever happens,
Hume's picture would be correct, of emotion followed later by rational
reflection.

But other conducive properties led Bryson to the equally powerful
emotions of beauty and of awe, and these do depend on those parts of the
neocortex dealing with the higher stages of sensory processing – of sight
('the most silencing sight that exists on earth') and sound ('fretful shiftings
of snow'); and here the *whole person* is involved (not just the brain), in
what Danto called, as the title of his book on the philosophy of aesthetics,
'the transfiguration of the commonplace'.[27] As Bryson went on:

> Nothing prepares you for the Grand Canyon. No matter how many times you
> read about it or see it pictured, it still takes your breath away. Your mind, un-
> able to deal with anything on this scale, just shuts down and for many long
> moments you are a human vacuum, without speech or breath, but just a deep,
> inexpressible awe that anything on this earth could be so vast, so beautiful, so
> silent.[28]

Body *and* brain, reason *and* emotion, are tied together, as they still are
when the experience is remembered later, particularly when it is recap-
tured in such things as music or poetry. The experience is remembered,
not only by being *thought* about, but also by drawing on emotional mark-
ers in the body. In other words, even memory is not just rational alone: it
involves the recapitulation of embodied feelings and emotions as well, as
it most certainly does in the most obvious case of sexual fantasy.

That is a major reason why, in the history of religions, what I have
called 'somatic exploration' and 'somatic exegesis' (from Greek, *soma*, 'a
body') are so important.[29] The highly differentiated forms of social organization and individual commitments, which have evoked the words
'religion' and 'religious', derive from the discoveries that were made in the
past, and continue to be so in the present. The integration of reason and
emotion is clearly fundamental in making both the exploration possible
and the exegeses shareable between people and through time.

But for that to happen, the consistency and independence of the
conducive properties are a necessary condition. At the level of the neurophysiology involved, they are independent of any particular culture or
tradition, though of course the opportunity offered by the conducive

properties leads to vast cultural and individual differences. *Maya* in India (see Note 11) is not the same as idealism in nineteenth-century Oxford, though both rely on the same underlying neurophysiology of perception. What makes the English laugh is notoriously different from what makes the Germans laugh, but because of the way that conducive properties invade them, they do both laugh. What turns people on sexually is clearly not the same in all cultures, nor even the same for all people in a single culture, but what the conducive properties lead to is recognizably similar in all people.

So yes, our aesthetic judgements are undoubtedly approximate, provisional, corrigible and often contested, as are those, in different ways, of scientists and historians. But they are nevertheless *judgements* based on the direct seeing and experiencing, both emotionally and rationally, of conducive properties, not as Hume supposed on perceptions giving rise to sentiments that are in turn qualified by reason.

So, while it is true that there is no single property of beauty that all must see, there are nevertheless many stable conducive properties that do lie in the objects or the people around us, and which evoke the judgements of satisfaction and value, and the vocabularies of beauty and goodness.[30]

They do so because the conduction, to give it that name, is so consistent. Think of the way that conducive properties in music reach people consistently: mood music and soul music, for example. Purcell's music for the funeral of Queen Mary is very different from Elgar's Pomp and Circumstance. The conducive properties in those two instances evoke the psychosomatic responses of grief or of patriotic pride.

There are, therefore, conducive properties which composers understand very well, and on which they can rely in order to create effect. That is why virtually the whole of the pop music industry has made so much money, because it is able to manipulate conducive properties with great skill. Hank Ballard, the R&B singer who invented the Twist in 1958, said just before he died, 'There's no medicine out there as good as music. ... If you're looking for longevity, just take a dose of rock and roll.'

In the case of art, the directness and consistency with which conducive properties can be seen is a major reason why Whistler could declare:

Art happens - no hovel is safe from it, no Prince may depend upon it, the vastest intelligence cannot bring it about, and puny efforts to make it universal end in quaint comedy and coarse farce. ... We have, then, but to wait - until, with the mark of the Gods upon him - there come among us again the chosen - who shall continue what has gone before. Satisfied that, even were he never to appear, the story of the beautiful is already complete - hewn in the marbles of the Parthenon – and broidered, with the birds, upon the fan of Hokusai – at the foot of Fujiyama.[31]

It is extravagant language but there is a truth in it. The story of the beautiful *is* already complete, however much it goes on being told in new ways, because the satisfaction involved in the neurophysiological process remains stable, no matter when or where we observe the world around us.

In that sense, ironically, the conducive property evoking the judgement of beauty actually does lie – or can lie – within the circle, as all Zen Buddhists know when they contemplate or create the *enso*,[32] and as the Papal emissaries to Giotto in the fourteenth century found out for themselves. Hume insisted that 'beauty is not a quality of the circle'. But conducive properties certainly *are* within the circle in such a way that they can lead to the emotion that evokes the judgement and the vocabulary of satisfaction, including beauty. When Pope Benedict XI sent out emissaries to find the best artist for St Peter's in Rome, Giotto drew for them only a circle. The Pope and the emissaries saw in it both skill and beauty.[33] The properties in the circle which evoked that judgement are not those in the *enso*: the *enso* is a deliberately broken circle, a conducive property leading into the recognition of the breaking of the otherwise endless circle of rebirth; the unbroken circle is endless, a conducive property leading into the recognition of the infinity of God – a property so much within the circle that it was able to be developed and exploited in subsequent art, literature and theology.[34]

This is the truly important conjunction in art, between skill and its competence to bring into being the conducive properties that evoke in the observer the emotion and the judgement of beauty or of other satisfaction (in the case of the *enso* of enlightenment, in the case of theology of contemplation) because it is this that creates a real distinction between art and mere artefact.

That, however, is exactly the distinction that many postmodernists are trying to deny or to destroy, at least in the practice of their own understanding of art. Etymologically, in the original meaning of the Latin word *ars, artis*, the postmodernist understanding happens to be correct. The word *ars* was used originally to describe skill in joining things together, or skill in undertaking some trade or work with one's hands, or with one's head – skill in speaking, for example. It was used to translate the Greek word *techne*, the word we use in terms like 'technology' or 'technique'. So we find in Latin many compound phrases like *ars duellica*, the art or skill of warfare, *ars musica*, *ars gymnastica*, and of course the title of Ovid's famous poem, *Ars Amatoria*, the technique or the skill of seduction. On that basis, Cicero, in the first century BCE, could quote Zeno, the Stoic philosopher, as saying that 'the true nature of art [artis proprium] is creare et gignere', 'to create and to bring to birth'.[35]

On that definition, the fundamental characteristic of art would be to bring something new into being outside the boundary of the human body. In that case, anything that humans produce outside the body is a work of art. On *that* definition, there is no difference between art and artefact, between Rembrandt and Tracey Emin's unmade bed. Indeed. And that is exactly why some contemporary artists can regard human excretions as a form of art, as, for example, in Piero Manzoni's 'Merda d'artista' of 1961. In the strict Latin sense, of bringing into being something outside the boundary of the human body, that is indeed art, as is virtually any production in the period that developed 'the shock of the new', followed rapidly by 'the shock of the now'.

But most of that is art only in the original Latin definition of the word, something produced outside the boundary of the human body. Under that definition, any artefact can be called a work of art: the two are simply equated. That is why, instead of the now-forbidden question, 'What is art?', we are encouraged to ask the entirely different question, 'When is art?'

What postmodernism does is give a completely unequivocal answer to that question, 'When is art?'. Postmodernism answers, 'Always': always when something is produced outside the boundary of the human body. If you ask a postmodernist, 'What is *not* art?', the answer is, 'Nothing', because the controlling conducive property is now confined to novelty and shock.

So the irony of *post*-modern art is that it has reverted to the most traditional and *pre*-modern definition of art, *artis proprium creare et gignere*: the true nature of art is to create and bring something into being that would not otherwise exist. But that is so vague that it is entirely uninformative: it includes in relation to humans virtually the whole of everything; and while that does indeed provide a rationale for postmodern art, and certainly gives rise to the Turner Prize each year, it has done so only by holding our humanity in contempt.

It does that because it refuses to recognize the truly important discriminations to be made in the case of art once we understand the neurophysiological way in which independent conducive properties generate appropriate and shareable vocabularies of beauty and goodness, in that human conjunction of seeing, emotion and rationality. This in its way completely deconstructs Derrida's own deconstructionist exhibition which he set up in the Louvre in Paris in 1990. It was based on Suvée's picture of Butades (illustrating the classical belief that the origins of art lie in the perfect copy of what the artist observes: this is discussed further on pp 56-7) in order to make the point that art is an expression and exploration of blindness. Indeed, the exhibition was entitled, *Les Mémoires d'Aveugles*, The Memoirs, but also - with a typical Derrida play on words - the Memories of the Blind.

Art is thus defined as an intervention between presence and absence, in the way that a holiday snapshot excludes virtually everything in front of a photographer: as the picture is taken, a frame is created for it, so that the photographer is blind to far more than is seen. When the developed photo is looked at, it creates memories that are not 'in' the scene at all, certainly not in the scene as anyone else might have observed it.

These notions of frame and blindness put paramount emphasis on the sovereignty of the subject - exactly the same kind of emphasis that has led to relativism and subjectivity everywhere else in the postmodern world. It means, inevitably, that all art is private, even when it is exhibited in public.

Exhibition shares a particular consequence of privacy. But can it establish something objective to which all people have equal access, some matters of fact to which appeal can be made, not least in the generalizing of at least some aesthetic judgements? Deconstruction and postmodernism answer, No, because post-Hume there is no single property of beauty to which, as a matter of fact, that appeal can be made. However, there certainly do exist conducive properties in the plural that do lead to consistent and shareable aesthetic judgements among organisms constructed in the way that humans are. These are the conducive properties which all whose eyes are open can see, and which the traditions of art both nurture and sustain.

For the fact is that conducive properties do invade the kind of creature that we are in very distinct ways. They evoke the different feelings and emotions that make us human, ranging from terror and hate to relief and love. Among them all, conducive properties certainly evoke the profound satisfactions that create and sustain the vocabularies of beauty.

The important truth is that we are not simply the passive recipients of conducive properties, as when a young bird cowers at the cardboard silhouette of a hawk. We both search for them and deliberately create them in the many different kinds of human creativity - in works, for example, of poetry or of music or of art. That is why Leonardo da Vinci encouraged would-be artists to look for them in otherwise uninformative surfaces:

I will not refrain from setting among these precepts a new aid to contemplation which although seemingly trivial and almost ridiculous, is none the less
of great utility in arousing the mind to various inventions. And this is, if you
look at any walls soiled with a variety of stains, stones with variegated patterns,
when you have to invent some location, you will therein be able to see a resemblance to various landscapes graced with mountains, rivers, rocks, trees,
plains, great valleys, and hills in many combinations. Or again you will be able
to see various battles and figures darting about, strange-looking faces and costumes, and an endless number of things which you can distil into finely-
rendered forms. And what happens with regard to such walls and variegated
stones is just as with the sound of bells, in whose peal you will find any name
or word you care to imagine.[36]

It is this that his immediate predecessor, Alberti, regarded as the origin of
art itself – an improbable speculation, but one that does not affect the
point about conducive properties:

I believe that the arts of those who attempt to create images and likenesses
from bodies produced by Nature, originated in the following way. They
probably occasionally observed in a tree trunk or a clod of earth and other
similar inanimate objects certain outlines in which, with slight alterations,
something very similar to the real faces of Nature was represented. They began
therefore, by diligently observing and studying such things, to try to see
whether they could not add, take away or otherwise supply whatever seemed
lacking to effect and complete the true likeness. So by correcting and refining
the lines and surfaces as the particular object required, they achieved their intention and at the same time experienced pleasure in doing so.[37]

Given, therefore, that artists, just as much as musicians, bring conducive
properties deliberately into their works, and given also, in the case of art,
that they notice them with educated eyes (a Christian will 'see' the cross in
a hedgerow where others will see only two branches), the question becomes inevitable: what *are* the conducive properties that lead to the
emotions of satisfaction and judgements of beauty?

It is relatively easy to list the most obvious. In the case of art, for example, many of them were gathered by Charles Johnson in his book *The
Language of Painting*,[38] a book that begins with the forthright sentence
(italics added), 'Painting is one means of expressing emotions *and* ideas.'[39]
But the property analysis required to understand how the conduction
leading to judgements of approval and disapproval works is far more
difficult. In the case of music, for example, Charles Rosen's analysis of the
classical style in Haydn, Mozart and Beethoven is in fact an analysis of
how different properties were combined under the constraint of particular
expectations of both form and tonality. Of tonality, he stated: 'The musi-

cal language which made the classical style possible is that of tonality, which was not a massive, immobile system but a living, gradually changing language from its beginning.'[40] Therefore he could argue:

> What unites Haydn, Mozart, and Beethoven is not personal contact or even mutual influence and interaction (although there was much of both), but their common understanding of the musical language which they did so much to formulate and to change.[41]

At the same time, there was an exploration of form, of, for example, what came to be called the sonata form, which again acted as a constraint (as will be seen later, pp 144-7, constraints are the condition of freedom which not only allow but demand creativity). The conducive properties in tonality and form are accessible, and therefore they are publicly and socially shared. Atonality is a step in the direction of the kind of elite privacy explored by Derrida (p 53). Innovation, however, does not depend on privacy of the restricted kind that seemed inevitable during the twentieth century. Thus Rosen commented on the first of Haydn's op. 33 quartets:

> We can see from this example how free classical form was, and how closely it was tied to tonal relationships. ... He [Haydn] is bound less by the practices of his contemporaries than by a sensitivity to harmonic implication; the suggestion of D major placed at a point as critical as the opening measure makes Haydn realize that he can dispense with a modulation. The same sensitivity will lead Mozart, after La Pinta Giardiniera, to write each opera in a definite tonality, beginning and ending with it, and organizing the sequence of numbers around it.[42]

In the case of art, property analysis is equally possible, but equally demanding.[43] The work of John Gage on colour is an example. It so happens that the properties of the spectrum of colour are independent of our opinion, but the ways in which humans might exploit the opportunities of those independent properties are virtually infinite. Nevertheless, they give rise to the cultural explorations of colour that are endorsed in social terms, because the underlying conducive properties are directly and consistently seen. Thus one of Gage's books[44] ends with the equally important sentence, 'The struggle to understand the nature of colour, whether physical or psychological, and to use that understanding in the shaping of our coloured environment has been the central subject of this book; it is a struggle that is still going on.'[45]

The struggle is unending, because it involves necessarily both reason and emotion as they are integrated, not just in idiosyncratic individuals, but also in social norms. How the physical opportunities of conducive properties are developed psychologically in different societies and cultures, and are then judged in terms of satisfaction and of the vocabularies of beauty, is exactly what the history of art reveals. The judgements are corrigible indeed. In 1929, the Marylebone magistrates described D.H. Lawrence's paintings of the nude as 'gross, coarse, hideous, unlovely and obscene'. In 1989, the Customs and Excise Department declared them fit for public display on the ground that 'the tide of public opinion has changed, and they would now be regarded as antique works of art.'[46] What endure are the conducive properties (along with, in terms of the underlying neurophysiology, the human ability to see them), but their consequence will be expressed very diversely in different times and in different cultures and societies.

This can be seen very briefly in the example of what in the West is called *mimesis*, since it (under different names) is found in both European and Chinese art, thus making it possible to see the way in which a conducive property relies on the same basic neurophysiology while at the same time producing immensely different consequences in each case.

Mimesis

Mimesis is a Greek word meaning roughly 'imitation', or more sharply in Western literature, 'the representation of reality', that being the subtitle of Erich Auerbach's famous book, *Mimesis*.[47] In literature it gave rise to the long-running 'sound and sense' debate: is it the case that words contain their own conducive properties, so that they convey to the reader the qualities or the emotions that a writer wishes to express? Alexander Pope, in An *Essay on Criticism*, made the strong claim that they do:

> 'Tis not enough no harshness gives offence,
> The sound must seem an echo to the sense.

Dr Johnson, in his *The Lives of the English Poets*, rejected this in a way that anticipates the argument of Hume about the imposition of sentiment on perception:

> The fancied resemblances, I fear, arise sometimes merely from the ambiguity of words; there is supposed to be some relation between a *soft* line and a *soft* couch, or between *hard* syllables and *hard* fortune. ... It may be suspected that in such resemblances the mind often governs the ear, and the sounds are estimated by their meaning.

In art, *mimesis*, the ability to represent – to re-present – reality has been a less contested conducive property, evoking, as it does, satisfaction and the vocabularies of approval. It appears in both the East and West, and in the West it was recognized as such very early on. In fact, it was once thought to be the origin of art itself. This can be seen in the paintings by Suvée or by Wright portraying the origin of art. Their paintings are based on the story of the daughter of Butades who has been told by her lover that he is about to depart for war. Desperate in love, she therefore stands him against a wall with a strong light shining on him, and she draws the outline of his figure on the wall to serve as a reminder of him while he is away.

When Pliny the Elder wrote his *Natural History* in the first century of the Christian era, he reported that no one could tell whether the Egyptians or the Greeks invented art, but at least, he wrote, 'all agree that art began with tracing an outline round a person's shadow'. In the Mediterranean world from which the great traditions of Western art are derived, one of the conducive properties leading to satisfaction is *mimesis*, imitation, and so it remained. Of a famous painting by Holman Hunt, known as 'Our English Coasts' or later as 'Strayed Sheep',[48] Ruskin wrote:

> It showed to us for the first time in the history of art, the absolutely faithful balances of colour and shade by which actual sunshine might be transported into a key in which the harmonies possible with material pigments should yet produce the same impressions upon the mind which were caused by the light itself.[49]

This is *mimesis* as a conducive property leading to satisfaction in the brain, and not just for Ruskin alone.

If, however, *mimesis* stood alone as the *only* conducive property, it would be a kind of surrogate for the single property of beauty. In that case, the highest form of art would be *trompe l'oeil*, paintings that trick the eye into supposing that it is looking at the scene or the object itself and not at a picture. Many artists have painted such pictures as an exercise. Giotto could do this so well that while he was still a boy he loved to paint a fly realistically on a portrait that his teacher, Cimabue, was painting, simply to see his teacher come back into the room and try in vain to brush the fly away.[50]

Clearly, therefore, *mimesis* on its own is not enough. For sure, it requires skill and technique, but even then artists, if they are to attain their purposes, will still be faced with the question of what to create using that conducive property in conjunction with others. The question preoccupied Rodin throughout his life, not least in considering the difference between an artist and a photographer. In 'My Testament', towards the end of his life, he made this distinction: 'Mere exactitude, of which photography and *moulage* [casting from life] are the lowest forms, does not inspire feelings.'[51] Yet he was capable of such mimetic exactitude that his first work to be exhibited (in 1877) caused a major scandal: he was accused of having cast from life the statue 'The Age of Bronze'. After the issue was taken to adjudication, one of the jurors remarked, 'Even if it is cast from nature, it's very beautiful. It should be accepted anyway.'[52]

For the juror in question, *mimesis* was a conducive property, leading to an aesthetic judgement of beauty. Even so, it is not enough on its own. The words 'mimesis' and 'beauty' are far from being synonyms. Thus when another public protest denounced Rodin's mimetic sculpture of an old prostitute,[53] on the grounds that it was grotesquely ugly, Rodin fired back:

> The vulgar readily imagine that what they consider ugly in existence is not a fit subject for the artist. They would like to forbid us to represent what displeases and offends them in nature. It is a great error on their part. What is commonly called *ugliness* in nature can in art become full of great beauty. ... In fact, in art, only that which has character is beautiful. Character is the essential truth of any natural object, whether ugly or beautiful. ... It is the soul, feelings, the ideas expressed by the features of a face, by the gestures and actions of a human being. ... Now to the great artist, everything in nature has character. ... And that which is considered ugly in nature often presents more character than that which is termed beautiful. ... In all deformity, in all decay, the inner truth shines forth more clearly than in features that are regular and healthy. ... Whatever is false, whatever is artificial, whatever seeks to be pretty rather than expressive, whatever is capricious and affected, whatever smiles without motive, bends or struts without cause, is mannered without reason; all that is without soul and without truth; all that is only a parody of beauty and grace; all, in short, that lies, is ugliness in art.[54]

Given, therefore, the power of *mimesis* as a conducive property leading to aesthetic judgements of beauty, any subject matter is possible, and its use will depend on the choice and decision of the artist – exactly as George Eliot realized when she made her plea that artists should 'love that other beauty too, which lies ... in the secret of deep human sympathy':

> All honour and reverence to the divine beauty of form! Let us cultivate it to the utmost in men, women, and children – in our gardens and in our houses. But let us love that other beauty too, which lies in no secret of proportion, but in the secret of deep human sympathy. Paint us an angel, if you can, with a floating violet robe, and a face paled by the celestial light; paint us yet oftener a Madonna, turning her mild face upward and opening her arms to welcome the divine glory; but do not impose on us any aesthetic rules which shall banish from the region of Art those old women scraping carrots with their work-worn hands, those heavy clowns taking holiday in a dingy pot-house, those rounded backs and stupid weather-beaten faces that have bent over the spade and done the rough work of the world – those homes with their tin pans,

their brown pitchers, their rough curs, and their clusters of onions. In this world there are so many of these common coarse people, who have no picturesque sentimental wretchedness! It is so needful we should remember their existence, else we may happen to leave them quite out of our religion and philosophy, and frame lofty theories which only fit a world of extremes. Therefore let Art always remind us of them; therefore let us always have men ready to give the loving pains of a life to the faithful representing of commonplace things.[55]

In contrast, Holman Hunt regarded all that as leading to the degradation of art:

Instead of adorable pictures of nature's face, we are offered representations of scenes that none but those with blunted feelings could contemplate, not stopping short of the interiors of slaughter-houses. The degradation of art is nothing less than a sign of disease in Society. But enough of this humiliating topic!

What did he offer instead? *Mimesis* still, but combined with other conducive properties - synecdoche, for example, and the demand made upon the observer of moral uplift. The passage above continues:

But enough of this humiliating topic! I must return to the defence of the Pre-Raphaelites. After fifty or sixty years, with full count of our disappointments as of our successes, it may be confidently affirmed that the principle of our reform in art was a sound one. With some remarkable exceptions, art in our youth had become puerile and doting, and it was high time to find a remedy. It stirred us to proclaim that art should interpret to people how much more beautiful the world is, not only in every natural form, but in every pure principle of human life, than they would without her aid deem it to be. If artists' work misguides men, making them believe that there is no order in creation, no wisdom in evolution, decrying the sublime influences as purposeless, we shall indeed be a sorry brood of men.[56]

Mimesis remains a conducive property that he valued and used, but he combined it with other conducive properties to work out his own pre-Raphaelite purpose. Conducive properties like *mimesis* belong to the reasons why a particular work of art comes into being, but so too does the purpose of the artist. Conducive properties are no guarantee that a particular work in which they are incorporated will evoke the aesthetic judgement of beauty. They *are* a guarantee that all people will see them and make their own judgements.

This means that conducive properties are available for all the enterprises of art, including art as propaganda. Because they are properties that lie within the object, they are extremely useful, as any advertiser or politician knows well. When rulers have wished to impress their subjects they have made artists paint them literally larger than life. The portraits of Henry VIII, derived from Holbein's mural which the Barber Surgeons Company commissioned him to paint in 1539, make the king dominant over all that he surveys. The version in Trinity College, Cambridge, uses *mimesis* as a conducive property, but only to evoke feelings of awe almost akin to worship. The original includes a verse that addresses Henry almost as a God:

> We, therefore, a suppliant band of thy Physicians
> Solemnly dedicate this house to thee,
> And mindful of the favour with which thou, O Henry, hast blest us
> Invoke the greatest blessings on thy rule.[57]

When the purpose of art is to influence people's minds, propaganda often uses realistic techniques, but in a way that in fact subverts reality. What, for example, did Queen Elizabeth I look like? Her portraits, like the poems about her, portray a myth. The portrait by Federigo Zuccaro in 1575 is about the only one that is thought to be like her, but it highlights the difficulty. The artist has made his sketch of the queen, but already to the right there is an allegory of the Queen's virtues, a column of Fortitude and Constancy, accompanied by the serpent of prudence and the dog of fidelity, and the ermine of purity on top.

At least that painting kept the person and her supposed virtues apart. Not so the portraits more usually painted of Elizabeth, on which Roy Strong has commented:

> It would be true to say that the Elizabethan royal portrait painter was con-
> cerned, not with portraying a likeness in the sense in which we should
> understand it, but with transforming a 'Rude counterfeit' into an image 'full
> of glory', an icon calculated to evoke in the eyes of the beholder those princi-
> ples for which the Queen and her government stood.[58]

By the end of her life, when the queen was well into her 60s, she was almost invariably portrayed without a wrinkle, as a young girl – as repeat-edly by Nicholas Hilliard in his miniatures. In a famous image, she is portrayed as one who has triumphed over both time and death.[59] This is *mimesis* of an image in such a way that it becomes, as Strong calls it, an icon – a reminder of the ways in which religious art so often uses the conducive property of *mimesis* for the purposes of propaganda and persua-sion.

Mimesis, therefore, persists as a conducive property, independent of our sentiments or opinions about it, in such a way that it is available to artists to use for their own purposes – as did Holman Hunt when he used it to raise the moral vision of the nation, believing that, as he put it, 'All art from the beginning served for the higher development of men's minds.'

That particular purpose was shared by many who used the conducive property of *mimesis* with great skill. It was shared, for example, by Benjamin Haydon (1786–1846), who used his mimetic skill in such well-known paintings as the very moving 'Blind Fiddler', the quiet humour of 'The Chairing of the Member', and the dramatic 'Raising of Lazarus'. But because he had that overriding purpose, he refused to paint simply for amusement, and as a result he spent virtually the whole of his working life owing money. On 11 June 1846, he listed all the people to whom he owed money – from Newton, £30, to the baker, £17.10.6 – 'in all £136. 14s. 10d., with only 18s in the house; nothing coming in, all [that is owed me] received'.[60] Eleven days later, he wrote in his journal a line from *King Lear*, 'Stretch me no longer on this rough world', he set up on an easel in front of him a portrait he had painted of his wife when she was young, he picked up a razor and he cut his throat.

Why, through unending poverty, did he persist as an artist? On 3 April 1812, he wrote: 'My canvas came home for Solomon, 12 feet 10 inches by 10 feet 10 inches – a grand size. God in heaven grant me strength of body and vigour of mind to cover it with excellence.' Even then he was broke: his friend Leigh Hunt (of whom Shelley wrote that he was 'One of those happy souls/Which are the salt of the earth, and without whom/The world would smell like what it is – a tomb'[61]) behaved nobly: 'He offered me always a plate at his table till Solomon was done.'[62] His landlord behaved even more nobly, forgoing rent until Haydon could afford it. As a result, wrote Haydon:

> I passed the night in solitary gratitude, and rising with the sun, relieved and happy, before setting my palette, prayed to the Great God who deserts not the oppressed, saying:

> 'O God Almighty, who so mercifully assisted me during my last picture, ... desert me not now. ... Enable me to conceive all the characters with the utmost possible acuteness and dignity, and execute them with the utmost possible greatness and power. ...

> 'O God, let me not die in debt. Grant that I may have the power to pay all with honour before Thou callest me hence. Grant this for Jesus Christ's sake. Amen.'[63]

He then commented immediately in his *Autobiography*:

Artists, who take up the art as an amusement or trade, will laugh heartily at this effusion of trust in God and this fear of being unworthy, but I took up the art by His inspiration. My object has ever been to refine the taste, to enlighten the understanding of the English people and make Art in its higher range a delightful mode of moral elevation.[64]

Using *mimesis* for that purpose did indeed excite the hearty laughter of Oscar Wilde when he declared, 'There is no such thing as a moral or an immoral book. Books are well written, or badly written. That is all.'[65] Substitute for the words 'book' and 'written' the words 'picture' and 'painted', and the result will be what has been, since the time of Wilde, a dominant assumption about art: the purpose of art is not to preach but simply to be – to adapt Archibald MacLeish on poetry.[66] According to the critic I.A. Richards, 'It is never what a poem *says* which matters, but what it *is*.'[67] Art for art's sake.

Yet still the fact remains that *mimesis* as a conducive property is independent of the purposes for which it is used. It is a property which people with the kind of brains that humans have see directly. Because the underlying neurophysiology is shared by us all, it is a fact about us on which the experience of beauty is dependent, exactly as Kenneth Clark insisted, long before any of this brain research was known about:

> There are very few people who have never had an aesthetic experience, either from the sound of a band or the sight of a sunset or the action of a horse. ... I believe that the majority of people really long to experience that moment of pure, disinterested, non-material satisfaction which causes them to ejaculate the word 'beautiful'; and since this experience can be obtained more reliably through works of art than through any other means, I believe that those of us who try to make works of art more accessible are not wasting our time.[68]

The universality of that experience does not depend on a property of beauty subsisting in objects of perception. But it does require a stable link between conducive properties and the underlying neurophysiology which enables us to see them in a way that combines emotion and reason.

So no matter how fundamental in Western art *mimesis* has been as a conducive property, it is never, outside *trompe l'oeil*, left on its own. *Mimesis* is not in any way operating as a substitute for a single property of beauty. Nevertheless, *mimesis* as the 'conveying through art the impression made by nature' acts as a conducive property, just as it does in other traditions of art.

To see this, we have only to look at China, where we find, not European *mimesis*, but a comparable emphasis in art on participation in the affective or stimulating images of nature, re-presented by the deliberate intention and skill of the artist. This was recognized so early in China as a conducive property leading to judgements of approval that it became one of the founding principles of art, just as *mimesis* did in the European tradition.

This we know, because it appears in the fifth-century work *Gu Hua-pin-lu* of Xie He, the basic primer of Chinese art which laid down the famous six principles. It is there, in two out of the six principles: first, *ying wu xiang xing*, 'responding by way of imitation to affective forms' (translated by Chiang Yee as 'modelling after object'), and second, 'transmission of the experience of the past' (translated by Chiang Yee as 'following and copying').[69]

But it is literally obvious, when one looks at a Chinese painting, that *mimesis* has been combined with so many other conducive properties that there is no intention on the part of any Chinese artist to re-present the scene as Ruskin believed that an artist should. Michael Sullivan began his book on the art of landscape painting in China by reproducing a painting in which three men are shown with shadows,[70] in order to make the point that this is the only painting he knows of from the classical period in China in which there are prominent shadows:

> How was it, we might ask, that the Chinese painter managed to convey so much without the aid of those pictorial devices that Western art through much of its history regarded as so essential? For the typical Chinese painting makes no use of scientific perspective, shading, or plastic three-dimensional modelling, and uses colour very sparingly or not at all. I have only found one instance of the use of cast shadows in the whole history of Chinese landscape painting. It occurs on a scroll illustrating Su Dong-po's *Second Red Cliff Ode*, where the poet describes how he and his friends stood one night on the river bank while the full moon threw their shadows on the ground behind them. All the artist can manage is a few smudges. But this is not pictorial realism as we understand it so much as a rather crude attempt to translate a vivid poetic image into a pictorial one. As we can see, it was not a success, and among the many later illustrations of this famous poem it never, so far as I can discover, occurs again; painters must have known instinctively that it was wrong.[71]

So Chinese painters did not make use of *mimesis* in order to anticipate photography, any more than Rodin did. Indeed, in that brief episode in the eighteenth century when the Jesuits Castiglione and Gherardini took Chinese names, they introduced Western techniques of realism into Chinese art. When the Chinese looked at their paintings, they were intrigued but also appalled. Offered a large architectural scene with pillars receding into the distance according to the principles of perspective, the Chinese touched the flat canvas in amazement. But they then said, 'There is nothing more contrary to nature than to represent distances where there are actually none, or where [on a flat piece of paper] they cannot be.'[72]

In that spirit, the artist Jing Hao (ninth/tenth century CE), who wrote a much-used guide called *Notes on Brushwork*, was immensely careful about the equivalent of *mimesis*. He recorded how he came across a valley in which pine trees were growing with astonishingly contorted forms – so much so that he went day after day to draw them until he had many thousands of sketches. But when he went to paint his finished works, he discarded the sketches altogether, because he knew that *mimesis* is only a conducive property, not the aim of the exercise. A young man once met him and claimed the opposite, that painting is to make beautiful things, and that the important point is to catch hold of their true likeness. Jing Hao replied:

> No, it is not. Painting is to paint, to estimate the shapes of things and really obtain them, to estimate the beauty of things and reach it, to estimate the reality of things and grasp it. One should not take outward beauty for reality. He who does not understand this mystery will not obtain the truth, even though his pictures may contain likeness.

When the young man asked him to explain the difference between likeness and truth, Jing Hao said:

> Likeness can be obtained by shapes without spirit, but when truth is reached, spirit and substance are both fully expressed. Those who try to express spirit through ornamental beauty will make bad things.[73]

So here is the crucial point: because humans share the same underlying neurophysiology, the same conducive properties will appear in many traditions, evoking satisfaction and vocabularies of approval. But different use will be made of them because they appear in entirely different contexts of opportunity (of materials, of patronage, of tradition, of education, of belief, and so on), and always they are modified by the way in which they are linked to other conducive properties – as they were by Holman Hunt when he combined *mimesis* with synecdoche and moral uplift.

So, with what other conducive properties was the equivalent of *mimesis* being combined in China to produce the recognizably distinct style of Chinese art? One, certainly, was the extent to which any finished work must express *both* feelings *and* ideas – not unlike the definition of modern conceptual art, in which (to quote the definition of Honour and Fleming) 'the artist's product is of less significance than the idea and process which brought it into being and of which it is only the record.'[74]

A Chinese version of this was stated as early as the eleventh century by Su Dong-bo (1037–1101), when he claimed that the purpose of painting is not to depict things as they might be seen. If you want *that* kind of work, he said, employ a professional painter. For a work of art to evoke the judgement of beauty, it must express the artist's own ideas and feelings. Michael Sullivan records him quoting a painter-friend who said, "'I write to express my mind and I paint to set forth my ideas, and that is all." Another painter of this time said, "I make paintings as a poet composes poems, simply to recite my feelings and nature."' [75]

If, then, Chinese art is dealing with ideas, it is not surprising that another conducive property in China is the extent to which artists have engaged in a conversation with the past, with their predecessors, one of the six principles (p 63). This is *mimesis*, not of the scene or of the person, but of the way in which great painters in the past had realized their vision and ideas.

Thus during the Yuan dynasty (1279–1368) many painters saw it as their responsibility to engage in what they called *fu gu*, 'restoring the past'. Zhao Meng-fu believed that 'only by restoring the spirit of the past (rather than merely the styles of the past) could art, like Chinese culture itself, recover its standing.' He is said to have written these words on a picture in the year 1301:

> The most precious quality in a painting is 'the spirit of antiquity', *gu i*. If it is not present, the work is not worth much, even though it is skilfully done. Nowadays, people who paint with a fine brush in a delicate manner and lay on strong and brilliant colours consider themselves skilful painters. They are absolutely ignorant of the fact that the works in which the spirit of antiquity is wanting are full of faults and not worth looking at. My paintings may seem to be quite simply and carelessly done, but true connoisseurs will realise that they are very close to the old models and may therefore be considered good. [76]

Note the word of approval, 'good'! And note also where he is said to have written those words: on a painting. In any art gallery in the West, that would be an act of vandalism leading to arrest. And yet in China, the 'writing on the picture' is a third, hugely important, conducive property leading to the judgement that a particular painting is or is not good. For it is obvious – again, literally obvious – that empty spaces in Chinese paintings are filled with writing and seals: the first three seals indicate the two names of the artist and some quintessential idea that belongs to the picture; the writing is usually the poetry that is part also of the picture, so much so that the calligraphy is often in the same style as the painting. In a sketch by Wu Zhen (1280–1354), the calligraphy extends the spiky upthrust of the pine needles, and the painting gives visual form to the poem which reads:

> To the west, the village of red leaves,
> The last light of the day is lingering.
> By the sandbanks of the golden reeds,
> The faint vestige of the moon just appears.
> So lightheartedly pull the oar, and let's return;
> Hang up the rod, and fish no more.[77]

It is because painting and poetry are two ways of saying the same thing that in China poetry is known as the host who issues the invitation and painting is known as the guest. The same Su Dong-bo who insisted that the purpose of his art was simply to express his ideas, wrote of Wang Wei (Wang Mo-jie): 'When I savour Mo-jie's poems, I find paintings in them; when I look at Mo-jie's paintings, there are poems in them.'[78]

So the successful integration of poetry, calligraphy and painting is a third conducive property which informs aesthetic judgement. A fourth is summed up in the word *li*, a word that pervades Chinese thought and society and eludes translation. Roughly it refers to all that unites people and society with each other and with the cosmos in appropriate behaviours with appropriate motives. Art is not exempt from that propriety. In his great chapter on *li* (Chapter 19), Xun Zi (in the work named after him) wrote:

> When *li* is at its best, human emotions and the sense of beauty are given full expression. ... It is through *li* that Heaven and earth are in balance, that the sun and the moon shine, that the four seasons are established in order, that the stars follow their path, that the rivers flow, that all things prosper, that love and hatred are regulated and that joy and anger are made appropriate. ... Those who hold to *li* are never bewildered even in the midst of greatest change, whereas those who wander away from *li* are lost. *Li*? Is it not the essence of culture?

Surpassing even that is a fifth property, the extent to which artists not only enter into the reality of what they are portraying in word and painting and become a part of it, but enable the onlooker to do so as well. According to the artists and scholars of the Song dynasty (twelfth to thirteenth centuries CE), the proper use of *mimesis* is for an artist, not to copy what lies before the eye or what others have done, but to interpret it through the character that has been formed through years of practice and thought. As Susan Bush summarizes the point:

Art presents a convincing image of the visible world; this image, when filtered
through the mind of an exceptional person, can be truer than nature uninter-
preted. This seems to be the message of the Sung critics.[79]

For those of them who were Daoists, the character and mind lend themselves to that kind of art. For them, the Dao is the unproduced Producer
of all that is, in itself far beyond word or description, but able to be recognized through its effects, through De. Daoist life is simply to realize and
identify oneself with the unfolding process of Dao in De, and thus to
recognize that one is an actual part of this process – there is indeed nothing else to be. For Buddhists (and many Chinese are both Buddhists and
Daoists), a work of art is to realize and be a part of the Buddha-nature (pp
186-7), of *śunyata*.[80] In both cases (and for many artists, Dao and *śunyata*
were just different ways of saying the same thing), a work of art is not
simply to contemplate, still less to portray, whatever the Dao or the Buddha-nature is: a work of art *is* the Dao and the Buddha-nature, since there
is nothing else. That is why Fu Cai could say of the work of Zhang Zao:

> When we contemplate Master Zhang's art, it is not painting, it is the very Dao
> itself. Whenever he was engaged in painting, one already knew that he had
> left mere skill far behind. His ideas reach into the dark mystery of things, and
> for him, things lay not in the physical senses, but in the spiritual part of his
> mind. And thus he was able to grasp them in his heart, and make his hand
> accord with it.[81]

It is because a work of art brings the artist into the final truth that one of
the supreme Daoist virtues becomes a conducive property in Chinese art,
the virtue of *wu-wei*. *Wu-wei* means roughly acting in the world but in such
a way that one is not acting, but simply allowing the Dao to form in and
through oneself. It can be seen in the story told about Zhen Hong-shou in
the seventeenth century, when (adopting the Chinese understanding of
mimesis) he copied a particular painting by Zhou Fang four times until
someone observed, 'Your painting is already far better that the original:
why do you go on copying it?' Zhen replied, 'My paintings look good
because of the skill that I use. Zhou's work is extremely skilful but seemingly without skill – and *that* is what is so difficult to achieve.'

Wu-wei applied to painting means that artists must not get in the way
of what is being produced through them. It is made clear, not so much by
what is put in, as by what is left out, by being suggestive, allusive, prompt-
ing – no shadows. When the Emperor Hui Song made painting a part of
the entrance exam for would-be court officials, he used to set a line of
poetry for the applicants to illustrate. One of these was, 'A monastery
buried deep in the mountains'. The winning painting was a scroll with

towering mountains but no monastery. In the foreground, however, was the tiny figure of a monk coming to a stream to fetch water: for the monk to be there, engaged in that task, the monastery can be inferred, though not actually seen. What is left out draws those who look at the painting, not just into the scene, but into the Dao that produced the scene.[82]

A landscape, therefore, is not something apart from the artist: it is the Dao, or the Buddha-nature, itself into which all people if they are wise (including artists) aspire to come. That is exactly what Sun Chuo expressed in his poem 'Wandering on Mount T'ien-t'ai':

> The Great Void, vast and wide, that knows no boundary
> Sets in cycle the mysterious Being, So-of-itself;
> Melting, it forms the rivers and waterways,
> Thickening, it turns into mountains and hills.[83]

The extent to which all this is realized in a painting (or for that matter a poem) is clearly yet another of the conducive properties that leads to a judgement that this is a good work. It is a reason why action painting appeared far earlier in China than in the West. During the Dang dynasty (618–906), there were painters who used their hair as the brush, or who would cover people with ink and roll them around on the silk or scroll of their painting. Much later, Gao Qi-pei grew his fourth finger nail long, then split it to make a pen closer to himself and his own participation in the picture. The wild immediacy of the action painters was converted into the actual process of Dao creating order and beauty out of seeming chaos. The purpose of the artist is harnessed to the controlling metaphors of Daoist belief.

Chinese art is recognizably and consistently different. Yet in China as much as in the West, artists stand before nature and transfer what they see to paper or other material. The results are indeed different, but the underlying neurophysiology that responds to conducive properties in what they see is the same. In terms of reason combined with emotion, it is a direct way of seeing that engages the satisfaction systems in the brain and body when the appropriate conducive properties invade our sensory systems.

The response is correspondingly both emotional *and* rational, and as a result it is felt, very often, through the whole body: you look at the Grand Canyon and it takes your breath away. Ironically, the pre-Humean languages, not least of the Bible, were entirely right: we have gut-feelings and heart-stopping moments – and as for what sticks in the stomach! It is, according to Carr's *The Dialect of Craven*, whatever remains in the memory with angry resentment.

It is thus of paramount importance in revising Henson's project (of knowing to what reason might appeal in discriminating among our competing judgements) that the manifestations of emotion through the body are remarkably stable: we laugh when we are amused, for example, and weep when we are in sorrow, in predictable ways. Music and pictures and words and much else besides have power to move those who hear or see or read them in ways that are both emotional *and* rational, exactly as A.E. Housman used to say of his shaving in the morning:

> Experience has taught me, when I am shaving of a morning, to keep watch over my thoughts, because, if a line of poetry strays into my memory, my skin bristles so that the razor ceases to act.[84]

It is not just Housman experiencing an emotion and then thinking about it: it is Housman, with his own biography, engaged in body, mind and memory in an emotional and rational way. It is exactly as Byron complained to Thomas Moore that he could never get people to understand that poetry is reason expressing excited passion: 'And who would ever *shave* themselves in such a state?'

It is equally the experience of Michael Tippett in America when he was taken to see some pictures painted by Shaker women. One of them he found so immediately striking that he was, as he put it, 'driven towards reconsideration of a perennial problem'. As he wrote at the time: 'At my first sight of it, before I had time to rationalize, I *thought* it might be a Tibetan mandala. I was deeply *moved*'[85] – reason and emotion cooperatively engaged.

Yes, rationalization may then deepen the experience, but it does not alter the way in which conducive properties have engaged the whole psychosomatic system in the first instance. Hume emphasized the second of those points, but missed the way in which reason is already engaged in the first encounter. In his essay 'Of the Standard of Taste', he recognized that people do make immediate responses to objects of perception, but because there is no single property of beauty, he had to conclude that those first and immediate responses may be full of confusion and error. They have to be brought by reason into the rational understanding of what counts as beautiful, founded on what he called, as we have seen, 'the common sentiments of human nature' (p 39). So he wrote:

> So advantageous is practice to the discernment of beauty that, before we can give judgment on any work of importance, it will even be requisite that that very individual performance be more than once perused by us and be surveyed in different lights with attention and deliberation. There is a flutter or hurry of thought which attends the first perusal of any piece and which confounds

the genuine sentiment of beauty. The relation of the parts is not discerned; the true characters of style are little distinguished; the several perfections and defects seem wrapped up in a species of confusion and present themselves indistinctly to the imagination. Not to mention that there is a species of beauty which, as it is florid and superficial, pleases at first, but being found incompatible with a just expression either of reason or passion, soon palls upon the taste and is then rejected with disdain, [or] at least rated at a much lower value.[86]

What we now understand better is the way in which rationalization is not usually divorced from the emotional experience, because reason is already involved in the initial perception, and that is why rationalization continues to bind reason and emotion together. Both are needed, even though we do not have always to call on both equally at all times in our appreciation of art, as Stuart Hampshire pointed out:

If in a museum I see, first, a sculptured figure from archaic Greece and, later, a Giacometti bronze figure, I may think them both magnificent works of art, even though the conceptions of art, and of visual representation, which they satisfy are radically different. I need to understand and to enter into both conceptions of visual representation in order to feel their excellence as works of art. But I can, and normally do, kick away the ladder by which I have climbed up to my aesthetic response. I do not need to say, 'This is a splendid work of art, given the conceptions of representation prevailing in archaic Greece or in post-Cubist, post-Surrealist Paris.'[87]

Now we can see exactly why Henson could stand in front of those portraits in Auckland Castle, and judge them to be depressing. There were particular conducive properties that invaded him and evoked his emotions *and* his reason in a single response. If his friend disagrees with him, there is an independent ground of appeal, insofar as they can specify which conducive properties have invaded them and have constrained their judgement. There *are* public reasons why they have formed the judgements that they have.

That is, in this respect, the importance of the recent work in neurophysiology, because the possibility of appeal to independent facts has not been destroyed as conclusively as we are often told – at least in the case of aesthetics.

But that can hardly happen, surely, in that other area to which Henson hoped to appeal, the whole field of ethics. There surely it still obtains that you cannot get values from facts, you cannot get an *ought* from an *is*, to quote the famous disjunction with which moral philosophy has lived for the last 250 years. But that exactly is what is now open to challenge, as we will see in the next chapter.

4 THE APPEAL TO VALUE: ETHICS AND HUMAN BEHAVIOUR

On 5 March 1848, Count von Eckstädt, the Secretary of the Saxon Legation in Vienna, wrote in his diary:

> A troubled, sinister mood prevails here in all circles. The Paris revolution has illuminated the obscurity of our position like a thunderbolt. ... It is to be foreseen with certainty that we shall wade through streams of blood.[1]

1848 is the year known conventionally as the year of revolutions, beginning in France and Austria. Would the revolutions cross the channel into England? There were many who claimed that if people would only attend to the lessons of history, they would be able to foresee with certainty that the British also will wade through blood.

That is a claim, not just about the future but even more about the past, a claim that we can read the past with such certainty that we can predict the future. It was an attitude that provoked Greville a week later, on 12 March of that same year, to make an entry in *his* diary on what he called the 'historical perplexities and the errors and untruths which crowd all history'. He wrote:

> I have always said that history is nothing but a series of conventional facts. There is no <u>absolute</u> truth in history; mankind arrives at probable results and conclusions in the best way it can, and by collecting and comparing evidence it settles down its ideas and its beliefs to a certain chain and course of events which it accepts as certain, and deals with it as if it were, because it must settle somewhere and on something, and because a tolerable and probable case is presented.[2]

It is a sharp summary of why Henson's belief that conflicts might be resolved by an appeal to history cannot be realized in the way that he hoped. He hoped that not only in the case of history, but also of aesthetics and ethics, it would be possible to identify certain and non-controversial facts that would act as an impartial judge in the many disputes of interpretation that divide people from one another.

That is the exact opposite of what Greville called in his diary *conventional* facts, i.e. facts that are agreed to be so by us. Conventional facts would clearly allow Henson's opponent, Charles Gore, to mobilize his

Anglo-Catholic party, whereas Henson hoped that the partisan differences between them might be resolved by appeal to *independent* historical facts.

That strategy was exactly the one that Henson did adopt in his disputes with Gore, and it remained important to him for the whole of his life. In 1926, looking back almost to the time he was at Oxford, he wrote in a letter to Lord Charnwood:

> Some 33 years have passed since I settled my opinion respecting the 4th Gospel in the phrase – 'It is not history, but the first and best commentary on the Gospel' – and there I stand still. I think we should be quite unable to conceive of an intelligible person in the Christ as this Evangelist pourtrays Him, if we had the 4th Gospel only; but with the historical Person already clearly presented by the Synoptics, we do find – or at least I do – in the narratives and discourses of the great Supplement an illuminating and satisfying Interpretation of the Person.[3]

Henson's belief that there are incontrovertible facts of history in the Synoptic Gospels in contrast to the interpretations of the Fourth was already being devastated by New Testament scholarship, as now equally by the claims of postmodernity that facts are never innocent of interpretation – or if they are, they cannot be interestingly so. Even if one allows the objectivity of evidence, and equates that with the facts to which Henson wished to appeal, the very selection of evidence, let alone the way it is handled, is itself a matter of interests.

In the case of aesthetics, the objectivity of discourse seemed even further removed because of the truth of Hume's argument that there is no single property of beauty contained in an object of perception which everyone who takes a look must see. However, the example of China and of the pre-Raphaelites made it clear that the post-Humean, postmodernist critique is far too strongly stated.

It is indeed the case that we do not see a single property called 'beauty', which is as much a part of the furniture of the world as are rocks and trees, whether we are here to see them or not. Nevertheless, we do see conducive properties that lead to emotion and satisfaction in the kind of brains and bodies that we happen to have, and some of them *are* a part of the furniture of the world (the 'golden, glittering contortions' of that piece of quartz, p 42), so much so that some at least lead to as consistent reactions in the case of animals as they do in the case of humans.

It is therefore on the basis of this direct seeing, in which reason and emotion are, in humans, working together, that the vocabularies of beauty have been formed. They are tied to facts about the objects and not only to facts about ourselves. There are conducive properties in the form of a snake that evoke the response of fear, and they do that not just in humans but in animals as well. And there are conducive properties in objects that evoke at least in humans the emotional and rational response of beauty. They belong to the reasons why we form judgements that can be appreciated by others. Maybe those who have not grown up in China or India need education to respond to Chinese art or Indian music (though even that is far from always being the case), but the fact remains that these and other forms of art *are* widely appreciated outside the traditions that formed and sustain them. To that extent, the conducive properties involved do offer a court of appeal in the case of disputed judgements.

But however much that works in the case of aesthetics and the history of art, it surely does not do so in that other area to which Henson hoped to appeal, the area of ethics and of moral judgement. For here we run into trouble: no matter how well that shift of focus works in the case of the feelings and emotions that evoke the vocabularies of beauty, it surely will not work in the case of ethical judgement and moral choice.

Ethics, Intuition and Emotion

Here surely the debates of moral philosophy during the twentieth century have established beyond contradiction the correctness of Hume's argument that we do not see a property of goodness but impose it on what we see. To return to his words already quoted:

> If we can depend upon any principle which we learn from philosophy, this, I think, may be considered as certain and undoubted, that there is nothing, in itself, valuable or despicable, desirable or hateful, beautiful or deformed; but that these attributes arise from the particular constitution and fabric of human sentiment and affection.[4]

Yes, of course, we respond emotionally to what goes on around us, but Hume's point is that there cannot be any unanimity of judgement in those matters because impartial reason does not enter into the matter by what he called a chain of argument and induction (pp 40-1), so that all rational people must agree, in the same way that all people in a good light and not colour-blind must agree that 'this book is red'.

Already we have arrived at G.E. Moore in Trinity College, Cambridge, holding up his hands in front of him and saying, almost in desperation, 'Here are two hands'. Why did he say that? In order to prove that he had knowledge of the external world, since any proposition embodying the sceptic's reason for doubting the truth of the proposition that G.E. Moore has two hands will certainly be less true than the proposition itself.

On the basis of that robust version of cognitive realism, Moore went on to refute any claim by sceptics to doubt moral knowledge. He did so by affirming the certain way in which goodness is simply known. But since the natural property of having two hands, or, to give his more usual example, the natural property of 'redness', can be observed directly in a way that what he called in consequence the non-natural property of 'goodness' cannot, Moore concluded that the non-natural property of goodness is simple and unanalysable, and must just be intuited to be known.

All this is familiar, but the point of referring to it is to serve as a reminder of the subsequent history of moral philosophy during the twentieth century through which that kind of intuitionism was called in question and alternatives to it were proposed, all of which led far away from any hope of identifying incorrigible moral facts. Why stay so closely in touch with a cognitive model if it leads to the fantastic scenario of special intuitive faculties discerning non-natural properties? Wiser, surely, to say that provided the gap between describing and evaluating is kept securely in sight, then value judgements are not describing anything: they express what a speaker approves of or disapproves of; or they are intended to evoke responsive feelings in the hearer; or they are intended to persuade the hearer to act in a way of which the speaker approves – to refer briefly to familiar versions of moral philosophy proposed in the twentieth century.

If, however, we return to Trinity College, none of that was any help to Bertrand Russell who, in a desperation almost equal to that of Moore, was heard in Great Court saying aloud to himself, 'There must be something more to saying that killing children is wrong than that I disapprove of it, and that I try to share my disapproval with others.'

But if we do not see directly properties like goodness or beauty in the same way that we see redness, so that all people not colour-blind must necessarily agree, and if, as Hume insisted, non-moral facts do not lead to moral judgements by way of necessary implication or entailment so that, once again, all sane people in a good light must agree, there surely has to be a radical disjunction between facts and values, more often expressed in the form, 'you cannot get an *ought* from an *is*, you cannot get a value from a fact'.

In that case, the hope of appealing to a natural morality that is observed by all people in all ages is doomed: the emotivists, or now the expressivists, would seem to be right in saying that there are no moral facts to be observed, that values are added on to our observations, and that our best hope of a moral world is to persuade others to share those evaluations.

But of course those Boo-Hurrah theories of ethics, as they were called, led directly into Russell's dilemma: surely there is something more to saying that Cinderella is beautiful and her two sisters ugly than that I approve of Cinderella, and that I invite others to share my attitude – the familiar exchange in the pantomime: 'Oh yes you do!'; 'Oh no we don't!'

If, however, no set of factual premises can ever conclusively entail an evaluative conclusion in the sense that for one statement to entail another, the first statement cannot consistently be accepted while the second is rejected, then the dilemma is inevitable: facts may be relevant to evaluation and moral judgement, and they may well help to resolve moral disagreements, but they cannot lead *coercively* to a conclusion or to an agreement. Facts may, but do not always for all people, lead to an evaluative conclusion.

The issue then becomes whether morality can make any claim to objectivity, exactly as Jonathan Lear put it in the opening paragraph of his essay, 'On Moral Objectivity':

> Morality exercises a deep and questionable influence on the way we live our lives. The influence is deep both because moral injunctions are embedded in our psyches long before we can reflect on their status, and because even after we become reflective agents, the question of how we should live our lives among others is intimately bound up with the more general question of how we should live our lives. The influence is questionable because morality pretends to a level of objectivity that it may not possess. Moral injunctions are meant to be binding on us in some way that is independent of the desires or preferences we may happen to have. When one asserts that a certain action is morally worthy or shameful one is, <u>prima facie</u>, doing more than merely expressing approval or disapproval or trying to get others to act as instruments of one's own will. If moral assertions were shown, at bottom, to be merely such exhortations, then they would be shown to wear a disguise. Morality would be revealed as pretending to an objectivity it does not have, and such a revelation could not but have a profound impact on our lives. It is doubtful that such a revelation could be kept locked up inside our studies.[5]

One might argue, as indeed R.M. Hare did argue against emotivists, that moral judgements are non-descriptive but that they are based, at least potentially, on rational evaluations which issue in universalizable prescriptions. Even so, that simply emphasizes the radical distinction between the observation of facts and the judgements of value. As Hare put it bluntly:

> The ascription of redness is governed by conventions which do not allow two people, faced with the same object in the same light in normal circumstances, to say, one of them that it is red and the other that it is not. One of them must be in breach of the conventions. He is in breach of them even if his mis-

take is due to colour-blindness. But the ascription of wrongness is governed by conventions which do allow you and me, confronted by the same act in normal and identical circumstances, to go on saying, one of us that it is wrong and the other that it is not wrong, if that is what we respectively think. We can reason about it in the hope that one will convince the other; but neither of us is constrained by our observation of the facts of the case and the correct use of words.[6]

In that case, it seems to follow that moral judgements cannot be true, because there is nothing for them to be true of. If moral judgements are matters of evaluative opinion, based on emotional responses (to which facts are relevant but not decisive), then clearly two people disagreeing about a moral issue are disagreeing about their own internal states of mind or attitude. If there are no moral facts, then there is nothing external to which two people in disagreement can appeal in order to resolve their disagreement, because *non*-moral facts cannot determine which attitude is right or wrong.

Not surprisingly, Bertrand Russell ended his life still passionate about his beliefs, but aware that they could not be true. He wrote as the opening words of his *Autobiography*:

> Three passions, simple but overwhelmingly strong, have governed my life: the longing for love, the search for knowledge, and unbearable pity for the suffering of mankind.[7]

But he could not see any way in which a moral proposition derived from those passions, those emotions, could be true. In 1944, he wrote, in a reply to his critics:

> An ethical judgment ought – so it is felt – to have the same kind of objectivity as a judgment of fact [exactly what Henson did feel]. A judgment of fact – so I hold – is capable of a property called 'truth', which it has or does not have quite independently of what anyone may think about it. ... I see no property, analogous to 'truth', that belongs or does not belong to an ethical judgment. This, it must be admitted, puts ethics in a different category from science.[8]

So it came about that the relativism of much moral philosophy of the last century, and certainly of postmodernism, has come to rest on three widely accepted claims:

- first, that there is no single property of goodness or evil that we are bound to observe in some act or person in the same way that we are bound to observe the redness of a book;

- second, that in consequence, we do not see directly good and evil, right and wrong, but add these judgements or evaluations onto our observations;
- third, that we cannot move from facts to value judgements because there is no logic of implication or entailment that will carry us over from one to the other in such a way that all people must agree in the way that they have to (if they are not colour blind) in the case of the redness of a book.

Two Extremes in Ethics

If those three claims are true, then a relativistic and non-cognitive position in ethics seems inevitable. Don Cupitt, in his book *The New Christian Ethics*,[9] took them so much for granted that he called those who take a different view (those who maintain that there are objective facts related to the values which humans affirm, and which indeed make them human) 'barmy',[10] ludicrously mistaken',[11] 'absurd'[12] and 'preposterous'.[13]

In contrast, Cupitt claimed that '*We* make the truth, we create reality, the knowing and the doing are one';[14] and again, 'It begins to look as if the whole notion of moral objects as being out there with some degree of reality or other, whether less or more, is a mystification.'[15]

The conclusion then follows (though it is in fact the opening of the book), 'that the ethical – just like the law, and like religion – is not sacred and timeless but a mere changing human improvisation.'[16] In contrast, he identified the position of those who support any kind of realism with what he calls 'Platonism':[17]

Realism staggers on because of the belief in the pure and primary datum, an objective and pre-linguistic Given that is better than and prior to our inadequate interpretations of it. They may fail, but it stands.[18]

But the argument here is (to use his own word) ludicrous, for two reasons. The first is the way in which realism is identified with Platonism (with the view that there is an existent, given idea of the Good, toward which human judgements inadequately aspire); Platonists are then identified as those for whom predications of good must be tracking heavenly archetypes. This does not allow that there may be other non-Platonic forms of realism – for example, the critical realism underlying much of this book.[19]

The second error is that this argument commits, in spectacular fashion, the fallacy of the falsely dichotomous question.[20] The fallacy is to say that morality depends *either* on tracking timeless, Platonic ideas and archetypes, *or* on human improvisation which pays attention only to scientific principles. The fallacy is to suppose that there are only two possible accounts of the matter, and that since the one is false, the other must be true. Realists (i.e. Platonists) must be wrong because, by refusing to acknowledge the only alternative, they fail to see that human creations, like law, religion and ethics, are a matter of improvisation. In Cupitt's view, we make them up as we go along. So he wrote:

> We may look back nostalgically to the days when design was traditional and traditionally just right, when spade and fork and drystone wall and five-bar gate were already exactly what they should be and did not have to be thought about. But in these days the most everyday articles need to be redesigned, consciously, rather often and on scientific principles. ... To suppose that there is some compulsory objective and timeless truth in these matters is like supposing that the design of a garden fork copies a heavenly archetype and cannot be altered. We just don't think like that any more.

It is much to be hoped that we don't think like that paragraph any more, because of the fallacy it contains. Nevertheless, it represents one extreme in ethics, the kind of subjectivism against which the other extreme in ethics, of objectivity, protests – as, for example, in many recent papal encyclicals.[21] In 1987, Graham Leonard (then bishop of London) called it 'the tyranny of subjectivism'. In November of that year he delivered the Green Lecture in America under that title, and argued that subjectivism is,

> based on the belief that principles and values are in essence no more than statements about the likes and dislikes, desires and aversions of those who hold them. From this it follows that there is no possibility of any resolution of disagreements about questions of principles and value or even of politics. ... There can be no more on the part of the subjectivist, than the restatement of what he considers to be self-evident truths and he is impervious to the arguments or criticisms of those who seek to challenge him.[22]

Leonard then made the claim that because subjectivism has no objective and independent standards of truth or morality outside the subject, it leads to violence and tyranny. He recognized that in the past, and in the ages of faith when the Church had absolute authority, bad things were done to good people, but at least, he argued, the Church in those days knew what it was doing and why. He therefore continued:

> It is for this reason that subjectivism leads to the use of violence to achieve its end, just as fundamentalism led to the Inquisition and to the rule of the saints in Calvinism, but with one significant difference. Fundamentalism could claim, albeit in the wrong way, to [be conforming to] the demands of truth outside man, whereas subjectivism leads to violence because it has no objective criteria to which it can appeal.

But this is both absurd and offensive. It is absurd because no example was given of subjectivism leading to tyrannical violence. In fact, the opposite is far more likely to be true. Those who say, 'do it my way or die', are almost invariably those who believe that they are objectively and absolutely right. As it was with the Church in an earlier age, when it operated in the style which encyclicals and the former bishop endeavour to continue at the present time, so it is with those who in the last century did most damage to immense populations of innocent people: Stalin, Hitler, Tojo, Pol Pot, the leaders of apartheid governments in South Africa, were not subjectivists. They had an absolute sense of being objectively right, and of conforming to natural laws, whether those of history or of race.

What is more, they all rejected subjective dissent with as much passion as the then bishop of London. Jews, homosexuals or women are more likely to be secure in the United States, which holds certain truths to be self-evident (the mark, above, of subjectivism), than they would be under the jurisdiction of the then bishop of London or of the Vatican. It is indeed the case that the Constitution of the United States is non-negotiable (although it is open to amendment), but to suppose that the United States is thereby committed to tyranny is absurd – the very country where moral protest led to the resignation of a president and the ending of a war. It is true that civil rights took a long and bloody time to achieve, but segregation was in the end abolished. And it is true that prisoners at Guantanamo Bay and in Iraq can hardly feel protected under a Bush/Blair administration that has (at least surreptitiously) abrogated the Geneva Convention; but here again the protests have been both vigorous and effective.

So Leonard's conclusion is absurd. It is also offensive, because it implies that there was something better about a Jew burning in the fires of Torquemada's Spain than in the ovens of Hitler's Germany, on the ground that at least the Inquisition was trying to adhere to truth and objectivity, external to itself. If this is where religious certainty leads, no wonder the world is marching to a different, non-cognitive step.

We seem, therefore, to be left with two extremes in ethics, irreconcilably divided from each other, neither of which is satisfactory. At one extreme lies the improvisation of morality of the kind advocated by Cupitt. At the other lies the claim that what is good and what is evil can be known objectively. At this extreme lies absolute authority in which morality is conformity. When associated with the commands of God, these commands are mediated through the institutions of religion (for example, of the Church, although in other religions they would be mediated through different authority systems, and not all the commands would be the same), and the proper response is obedience. What is more, that response of obedience might well be looked for universally, because the commands are claimed to be those of God. An appeal may therefore be made to natural law or to 'creation principles': natural laws like gravity apply to everyone, even though some may try to live in disregard of them by attempting to fly without wings. So too, on this argument, do the natural laws of ethics.

Thus natural law in Roman Catholicism is understood as the participation of human beings in the eternal law of God. It is therefore the guarantee of 'the objective norms of morality'.[23] There have been many different understandings of 'natural law',[24] but the one underlying Catholic pronouncements, for example in encyclicals, is that of Aquinas:

> Since all things subject to divine providence are ruled and measured by the eternal law, it is evident that all things partake in some way in the eternal law, in so far as, namely, from its being imprinted on them, they derive their respective inclinations to their proper acts and ends. Now among all others, the rational creature is subject to divine providence in a more excellent way, in so far as it itself partakes of a share of providence, by being provident both for itself and for others. Therefore it has a share of the eternal reason, whereby it has a natural inclination to its proper act and end; and this participation of the eternal law in the rational creature is called the natural law. Hence the Psalmist, after saying, 'Offer up the sacrifice of justice', as though someone asked what the works of justice are, adds, 'Many say, Who shows us good things?' [Ps. iv. 6], in answer to which question he says: 'The light of Your countenance, O Lord, is signed upon us'. He thus implies that the light of natural reason, whereby we discern what is good and what is evil, which is the function of the natural law, is nothing else than an imprint on us of the divine light. It is therefore evident that the natural law is nothing else than the rational creature's participation of the eternal law.[25]

In the encyclical *Veritatis Splendor*, this foundation of objective morality is summarized:

> God himself loves and cares, in the most literal and basic sense, for all creation (cf. *Wisdom* 7:22; 8:11). But God provides for man differently from the way in which he provides for beings which are not persons. He cares for man not 'from without', through the laws of physical nature, but 'from within', through reason, which, by its natural knowledge of God's eternal law, is consequently able to show man the right direction to take in his free actions. In this way God calls man to participate in his own providence, since he desires to guide the world – not only the world of nature but also the world of human persons – through man himself, through man's reasonable and responsible care. The *natural law* enters here as the human expression of God's eternal law.[26]

On this understanding of objectivity in moral judgements, it is possible to construct coherent moral systems. In this case, it rests on the assumption that the commands and prohibitions of God have been revealed with sufficient clarity to issue in authoritative instruction (as, for example, in encyclicals), backed up by promise and threat. Dick Westley, writing as a committed Catholic, remembered the consequence of this when he was growing up:

> I almost want to laugh at the thought of it now. It seems so completely foreign to what I have come to know about God and how he works in the world that I find it difficult to believe I actually lived in such mortal fear of him. But I did. So did most other adolescent boys with whom I went to high school. One could hardly avoid being scared to death at what might happen if he indulged in or enjoyed, ever so slightly, an erotic pleasure! Since we were students in all-male Jesuit high school and were not married, to indulge in *any* erotic (in those days called 'venereal') pleasure meant we had sinned mortally. If we died after such an episode before we got to confession or made a perfect act of contrition, it was curtains for us. I still have an oft-quoted scriptural text in my head from those supposedly 'carefree' days of youth: 'But God said to him, "You fool! This very night your life shall be required of you"' (Luke 12.20). I suppose that is why non-Catholic adolescent boys always derived more pleasure from their erotic episodes than their Catholic counterparts. Our pleasure was always marred by the thought that we were playing Russian roulette with our very lives. If the church wanted to make erotic pleasures less pleasurable, it succeeded.[27]

In strong systems of this kind (another example is that of Sharia-based Islam), it is well recognized that general commands have to be related to particular issues, hence the proliferation of case studies (of casuistry; for the schools of application in Islam see pp 137–8). However, the overall purpose of moral philosophy is to find in revelation and in natural law a ground abstracted from the relative and contingent as a foundation for moral life and judgement.

However, the problems in constructing that kind of system are well known. It depends, for example, on accepting that God exists, that God is the loving and caring source and sustenance of all creation (since otherwise the connection between natural law and eternal law collapses), and that whatever is claimed as revelation (for example the Quran or the Bible) does indeed come from God. Aquinas felt that he had shown by argument *that* God exists, and could know from revelation *what* God is like and what God has done.

However, by no means all people have accepted the validity of those arguments, nor are they self-evident truths, yet without them this form of

the argument for objectivity fails. On the other hand, for those within any
system making these claims, they are beyond argument. The moral judgements deduced from revelation and natural law are absolute and beyond
negotiation, just as the law of gravity is beyond negotiation for someone
who has stepped off a roof – hence the confidence with which Papal
encyclicals can state what is objectively right and wrong.

Even then problems remain in making direct applications to life of
absolute commands and prohibitions. The most searching problem in
practice is that of distinguishing between, on the one hand, commands of
a general kind that are not restricted in any way (these are known as
'context-independent commands') and, on the other, the application of
such commands to particular issues or occasions (these are known as
'context-dependent applications', though they often also take the form of
commands).

Thus the book of Leviticus records the context-independent command,
'Be holy as I am holy' (19.2). But what does it actually mean in specific
contexts of life to be holy? Jewish tradition, especially in Mishnah and
Talmud, is a record of the application of that context-independent command to particular issues and questions. Jews are commanded to keep the
Sabbath day holy, but what does that mean in practice, in one context
after another? The context-dependent applications of the command to
keep the Sabbath day holy rapidly became so numerous that they were
described as 'a mountain hanging from a hair'. For Orthodox Jews, those
context-dependent applications are treated as a part of the context-
independent command (they are held to be the orally transmitted, in
addition to the written, Torah), so that they have the force of authoritative
commands. For other Jews, especially Reform and Liberal Jews, they have
authority, but not necessarily in the form of commands that must be
obeyed.

The distinction between context-independent commands and context-
dependent applications, and the different ways in which that distinction is
handled, have created within other religions as well deep and often inimi-
cal divisions, particularly when context-dependent applications are
converted into context-independent commands. When Jesus was challenged (as other Jewish teachers were) to give his summary of the entire
Law, he replied, 'Love God, and love your neighbour as yourself.' That is a
context-independent command. The writings of the New Testament,
particularly the letters of Paul, show how that context-independent command was applied to the questions and issues of particular contexts. Deep
divisions and conflicts among Christians arise when some turn the context-dependent applications of those writings into context-independent

commands, as though the instruction that women should wear hats in church is a command of that kind – or, to give a far more serious example, when the prohibition on homosexual acts is converted from the context-dependent application that it is into a context-independent command.

The result of converting context-dependent applications into context-independent commands is a major reason why there is so much intransigent conflict among those who belong ostensibly to the same religion. In the Quran, there is a context-independent command that women should lower their gaze and guard their modesty, and should not display their beauty except to their immediate family (24.31), and that they should cast their outer garments (*jalabib*) over themselves, to be known for who they are and to avoid molestation (33.59). But in what exact ways should those context-independent commands be applied? In 2004, a school in England with a majority of Muslim students agreed with Muslim advisers (in a context-dependent application) on a uniform that met the requirements of the context-independent command of the Quran. However, one pupil applied the context-independent command of the Quran to herself in a way that required more in the way of covering than the agreed dress. In other words, she converted her own context-dependent application into a context-independent command, and the case ended up in the High Court (where the judgement upheld the majority application and rejected the pupil's interpretation).

One of the consequences of this conversion of context-dependent applications into context-independent commands is non-negotiable bigotry and hatred. Even when it operates on a less strident and destructive scale, it speaks with a far greater certainty than is possible on many ethical judgements. There are greater certainties in mathematics than in physics, much greater certainties in physics than in sociology, some greater certainties in sociology than in morals. The error is to claim for morals the certainties of mathematics, and then to dictate those claimed certainties to the world at large, as though every certainty can be solved by writing in and asking for the answer:

> Dear Miss Manners: Some time ago, a lady was dancing with her male friend at the White House, and her underslip dropped to the dance floor, and the lady just kept dancing as if nothing had happened. Was this the proper thing for the lady to do?

> Gentle Reader: Yes, the thing to do is to ignore it. A general rule of etiquette is that one apologises for the unfortunate occurrence, but the unthinkable is unmentionable.

Is the Bible Miss Manners? Or rather, are the clergy and the popes the Miss Manners and the agony aunts of the religious world? Clearly not. Moral life is not so simple, and of course religions laying claim to the revealed word of God know this and have built up impressive procedures and traditions of casuistry (the application of the law to individual cases). Thus even the Vatican backs away from its absolute judgements. It insists in *Veritatis Splendor* on the objectivity of the moral law in general and the absoluteness of moral prohibitions regarding specific human acts (§104), and on the latter it states, 'The Church has always taught that one may never choose kinds of behaviour prohibited by the moral commandments expressed in negative form in the Old and New Testaments' (§52). Yet the Church has obviously chosen the behaviour prohibited by the command, 'You shall not kill', by choosing war in some circumstances, and the execution of criminals.

The distinction, therefore, between context-independent commands and context-dependent applications is vital, and it is reflected in casuistry. But when we look around at the ocean of human disorder, the temptation is strong to unloose the knots that bind us to the mast of moral serious-ness and to give in to those siren voices on the shore, those voices of absolute certainty and determination, which remove from people respon-sibility and require of them obedience. Command and consent moralities resemble Mrs Pardiggle in *Bleak House* who 'took the whole [of the brick-maker's] family into custody – I mean into religious custody, of course; but she really did it as if she were an inexorable moral Policeman carrying them all off to a station-house.' It is true that 'Ada and I [Edith Summer-son] were very uncomfortable ..., and we both thought that Mrs Pardiggle would have got on infinitely better, if she had not had such a mechanical way of taking possession of people.' But in the religious world, there are many who are not in the least bit uncomfortable about acting as custodi-ans of other people's lives and morality. They may well 'look back in anger' at the Enlightenment as the age when their own authority began to be threatened (pp 10–11). In contrast, they look back in nostalgia to a Golden Age before 'the spiritual and, in particular, the moral patrimony of Christianity were ... torn from their evangelical foundation. In order to restore Christianity to its full vitality, it is essential that these return to that foundation.'[28]

A Golden Age. However, the last chapter of Kenneth Grahame's *The Golden Age* is called 'Lusisti Satis', you have played enough. And as Ed-ward disappears in the train that is carrying him for the first time to school, 'he was borne away in the dying rumble, out of our placid backwa-ter, out into the busy world of rubs and knocks and competition, out into

the New Life.'[29] But the New Life, at least as the New Testament understands it, does not suggest that moral doubt can be solved by writing in to
a bishop and asking for the answer.

These, then, are two extremes of moral judgement. At one extreme we
are offered the kind of authority that seeks obedience. At the other extreme, we are offered non-cognitive irrealism of the kind illustrated by
Cupitt, in which morality is an improvisation. That view is a natural
consequence of those three widely accepted claims listed above (pp 76–7).
But in my view two of the three claims are wrong.

What Leads to Judgement? Conducive Properties and Moral Judgement

The first claim, that there is no single property of goodness or evil that we
are bound to observe in some act or person in the same way that we are
bound to observe the redness of a book, is clearly correct. But the second
claim, that we do not see *any* properties directly that lead consistently to
judgements of good and evil is wrong, and wrong for the same reasons in
the case of moral judgements as in the case of aesthetic judgements: just as
we observe conducive properties that evoke and lead to judgements and
therefore also to the vocabularies of beauty, so also the same happens in
the case of goodness.

With the kind of brains and bodies that we happen to have, we clearly
do see conducive properties, in acts and objects and people, which evoke
feelings and judgements of approval and disapproval, and which thus
sustain vocabularies of good and evil, of right and wrong; and with these
brains and bodies, the responses we make are both emotional *and* rational
at one and the same time. Those who walked into Belsen in 1945 saw
immediately and directly what they knew to be evil. It was emotional, for
sure. But they had reasons for that judgement which were engaged at the
same time. Certainly they did not observe something which stirred their
emotions, to which later they brought rational reflection in order to arrive
at the judgement that this was evil. Martha Gellhorn was a frontline war
correspondent who had seen many horrifying things, but she had never
wept until she entered Dachau. She walked into Dachau and there, as she
put it, she fell over a cliff:

> Something changed for Martha that day; something to do with what she felt
> about memory and the past, and her own sense of optimism, and perhaps
> even about being Jewish. It was in Dachau, she said, that she really under-
> stood for the first time the true evil of man. 'A darkness entered my spirit,'
> she explained later to Hortense Flexner, and 'there, in that place in the sunny
> early days of May 1945' she stopped being young. 'I do not really hope now,'
> she wrote. 'Not really; I only feel one can never give up.' When, after a few

days, she fled Dachau, unable to bear the weight of her feelings any longer, she had lost her belief, carefully protected and nurtured – against all odds through wars and the Depression – in the perfectibility of man, the instinctive certainty she had clung to all her life that truth, justice and kindness would always, in the end, prevail. She no longer felt they would; and she never did again.[30]

Reactions of that kind are an immediate and integrated response to the direct seeing of conducive properties, leading to the formation of judgements which are in principle open to correction, but frequently turn out not to be so. The reasons why this is so go back to that same work on emotions and rationality which proved so helpful in understanding aesthetic judgements. Rolls, in his book *The Brain and Emotion*, made the same point twice when he wrote, 'Perhaps the most important issue in emotion is why only some stimuli give rise to emotion.'[31] Equally important is the issue why particular stimuli give rise consistently to one emotion rather than another.

The key to the answer, I am suggesting, lies in the fact that we do perceive conducive properties directly, in ways which, because they are both emotional and rational, lead to the different values associated with them. If, as seems to be the case, the reinforcement value of a visual stimulus is decoded in both the orbitofrontal cortex and the amygdala (pp 43–8), both these parts of the brain will produce outputs that constrain our behaviour.

An example of this direct seeing of conducive properties can be found in the work of Calder and others on facial emotion recognition after bilateral amygdala damage.[32] Calder and his colleagues studied two people with amygdala damage, in order to find out how much their ability to recognize facial expressions of emotion had been impaired. This they did using the Ekman–Friesen computer-manipulated hexagon of faces[33] expressing happiness and surprise, surprise and fear, fear and sadness, sadness and disgust, disgust and anger, anger and happiness. They then compared the speed and accuracy with which the amygdala-damaged people and people in a control group connected each image with the appropriate emotion.

The recognition of happiness, surprise and sadness was only slightly impaired in those with amygdala damage, disgust was difficult, and anger and fear were severely impaired. This clearly supports the argument (p 43) that more than the amygdala is involved, and also that while the amygdala may be more involved in the evaluation of some emotions than of others, it is not just involved in the emotion of fear – a reminder of the criticism that Rolls made of LeDoux (pp 45–6). Even more to the point, this research also makes it clear that in unimpaired people there is a direct recognition of conducive properties that engage evaluation.

The work of Scott and others[34] takes that point even further. They studied a woman with brain lesions confined to the amygdala region, and they found that the deprivation was not just a matter of visual recognition: she was also very poor at recognizing vocal expressions of fear and anger.

This means that the associated structures (including the amygdala) which are involved in our emotional responses, are involved in a general way, irrespective of the mode of input. That is why composers of music can so easily reach and evoke our emotions (pp 50, 114). So too can directors of films and producers of plays, who often use both sound and sight to create the conducive properties that evoke predictable responses in the target audience. They know exactly what the conducive properties are that lead to appropriate and common responses - fear, for example, laughter, or even disgust. Simon Moore's adaptation for the stage of a Stephen King story was reviewed in *The Independent* under the headline, 'Things That Make You Go "Argh"'. The review began:

> When two people fainted at last Saturday's preview of his new show, the writer and director Simon Moore was uncertain how to react. 'A part of me was alarmed; another part of me was thinking "Great!"'. This may sound a little callous, until you learn that Moore's play is an adaptation of Stephen King's thriller *Misery* [which includes a scene where 'Annie expresses her discontent by taking an axe to Sheldson's foot']. 'There's much less violence in the play than there is in the book,' says Moore, 'principally because it's very hard to stage violence in a convincing way. Also, I think the reason we have much tougher censorship laws on stage than in film or books is that when you stage things they move into a different area - it is more disturbing to a lot of people.' Deborah Warner has staged the gruesome blinding of Gloucester in *King Lear* twice. ... 'In the first one Cornwall leapt on to him and pinned him down - I think I wanted to focus on his enormous rush of energy and power. The second was the opposite, done with cool timing and scalpel accuracy. I don't think either was right. I now think what should happen is this: Gloucester's spectacles are mentioned very early in the play - Cornwall should stand on the glasses and stamp them into Gloucester's eyes. I wish I'd done that!'[35]

The conducive properties are so obvious that they have been claimed to offer humans evolutionary advantage in terms of natural selection and survival. Under the headline 'Emotions Keep Us Disgustingly Healthy', *The Daily Telegraph* made this report:

> The emotion of disgust evolved to help to protect us from infectious disease, according to scientists. The origin of our emotions has long been a source of speculation and the reason things such as bodily fluids, cannibalism and in-

cest make us go 'yuck' has long been debated. Today, in the Royal Society journal *Biology Letters*, scientists provide the best evidence yet that disgust is a protective emotion, as is fear of spiders and snakes, that helps to cut our risk of infection. Doctors from the London School of Hygiene and Tropical Medicine conducted a web-based experiment in which 40,000 people reported how disgusting they found a series of pictures including body lesions, faeces, rotting meat and parasites. Disease-relevant objects, such as a towel apparently stained with bodily fluids, were more disgusting than objects with no obvious disease association, such as a towel with a blue stain.[36]

The social construction of disgust extends the emotion far beyond that point of origin. Nevertheless, the basic fact remains: the recognition of conducive properties leading to judgements of approval or disapproval is fundamental in the kind of bodies that humans have – so much so that the recognition of what counts as good in relation to human good can be as direct as the recognition of what counts as good in the case of a stone wall or a fork. It does not belong to human good to jump on someone's eyes and grind their glasses into their eyes.

Natural Goodness

This point of view is much more Aristotelian than Platonic. It allowed the philosopher Elizabeth Anscombe to point out that there are facts about human life that are necessary if that life is to be possible, and which can therefore be called 'Aristotelian necessities' since good depends upon them: 'In ancient times, Aristotle in his *Metaphysica*, made the pregnant remark that one sense of necessary is *that without which some good will not be obtained or some evil averted.*'[37] From the point of view of Aristotelian necessity, it is clearly correct, rational and intelligible, not barmy, ludicrous and mistaken, to talk about a good fork or a good stone wall (p 78). When Youth Employment schemes set teenagers to work on rebuilding the stone walls of north-west England, the young people so employed soon realized that there is more to building such walls than merely piling one stone on top of another. There is a craft and a tradition acquired and passed on through time, as well as being limited by independent facts. Despite what Cupitt says, there are not many new ways in which you can improvise a new design for a dry stone wall, so that it still evokes the recognition that it is good.

Certainly it is not an infallible predication, nor is it incorrigible: the fork may bend and the wall may fall down. But we know what will count as a good fork and a good wall, not because they are pale shadows of ideal Platonic types, but because they perform their appropriate task well; and they are able to do that because factual limits are set on their possibility:

limits are set by what is in fact the case, in a world in which sheep may safely graze and earth needs to be turned over.

Equally, we know how to use the word 'good' in the case of plants and animals where the word 'good' is tied to the flourishing of the kind of organism that they happen to be. The structure of the argument is exactly the same for humans, even though it is, notoriously, far more complicated, to say the very least. As Philippa Foot has put it, 'The grounding of a moral argument is ultimately in facts about human life,'[38] i.e. facts which lie non-negotiably in the circumstances and conditions of human existence, and which therefore constitute what she calls 'a natural normativity'. The constitutive conditions of human life are manifest in conducive properties. We see them directly in a way in which feelings, emotions and rationality are at work together, and in which the judgements and the vocabularies of good and evil are evoked. At fundamental points of what belongs to the good of individuals and of social groups, we do have competence in judging what count as good lives and good acts in particular contexts, even though the predication is not infallible, nor is it incorrigible.

To give an example: in Anthony Trollope's novel *Framley Parsonage*, Lord Lufton tells his mother that he intends to marry Lucy Robarts and asks, 'Have you any objection to her?' Lady Lufton at once produces in her own mind a long list of conducive properties leading to the judgement that since she did not observe them in Lucy, she had every objection to her:

> For a moment or two Lady Lufton sat silent, collecting her thoughts. She thought that there was very great objection to Lucy Robarts, regarding her as the possible future Lady Lufton. She could hardly have stated all her reasons, but they were very cogent. Lucy Robarts had, in her eyes, neither beauty, nor style, nor manner, nor even the education which was desirable. Lady Lufton was not herself a worldly woman. She was almost as far removed from being so as a woman could be in her position. But, nevertheless, there were certain worldly attributes which she regarded as essential to the character of any young lady who might be considered fit to take the place which she herself had so long filled. It was her desire in looking for a wife for her son to combine these with certain moral excellences which she regarded as equally essential. Lucy Robarts might have the moral excellences, or she might not; but as to the other attributes Lady Lufton regarded her as altogether deficient.

The properties she listed are no doubt a social construction, belonging to a particular time, but again, that does not alter the way in which conducive properties are perceived, and in which they evoke judgements.

That is why those whom we might call cognitive or critical or (less elegantly) phenomenological realists are not in the least tracking Platonic Ideas lurking in some unobservable background. After all, Greek philosophers as early as the second century BCE had realized that, as David Sedley has put it:

> most of the epistemic jobs of Platonic Forms are discharged, without metaphysical complication, by *prolepsis*, the generic conception of a thing, compounded out of individual sensations of it.[39]

In other words, if a label *is* required, cognitive realists are more likely to be endorsing a corrected version of Aristotelianism, identifying what belongs to the good of any candidate for evaluation.

What certainly does not count as an Aristotelian good is a declaration that moral judgements are so subjective and relative that whatever at least some people claim to be good cannot be denied them – a point that became very clear when a Heavy Metal musician, Michael Cooke, allowed the inclusion of one of his songs in a compilation album, provided the record sleeve should not bear words or images 'associated with the world of black magic, witchcraft, satanic cults or the occult'. But the inside cover of the sleeve reprinted a review of one of the other bands, saying: 'It is the sound of a succubus mating orgasmically with a mortal man. It's the sound of Lucifer's voice emanating from Linda Blair's luscious lips. It's the sound of a sacrificial knife slicing bloodily between a virgin's pert little breasts.' The producer resisted demands that the record should be withdrawn, saying, 'It would be depriving kids of their musical heritage – this album is the celebration of a musical movement. It would be very unfair if it was consigned to the shelves just because of the maniacal obsession of one man.'[40]

Unfair or not, the objection to that flows, not from an irrational opinion, still less from a maniacal obsession, but from a fact-based objectivity rooted in the way in which we are constituted as human beings. What are factual are the conducive properties which evoke feelings and judgements of satisfaction and dissatisfaction, of approval and disapproval and which sustain the vocabularies of goodness and evil. What also are factual are the ways in which we are built that enable us to see them.

Thus conducive properties in relation to ethics are closely linked to Aristotelian necessities. This again is easy to see in the case of films. In the films of *The Lord of the Rings*, the conducive properties were poured in by the truck-load to make sure that the discernment of evil is, if not *absolutely*, then unmistakably, clear. Tolkien in the book was more nuanced, because he wanted to stay closer to the world that people experience (pp 129–30), but he knew the reasons why particular properties lead to the judgements that they do. Thus he wrote in a letter:

> In my story I do not deal in Absolute Evil. I do not think there is such a thing, since that is Zero. I do not think that at any rate any 'rational being' is wholly evil. ... In my story Sauron represents as near an approach to the wholly evil will as is possible. He had gone the way of all tyrants: beginning well, at least on the level that while desiring to order all things according to his own wisdom he still at first considered the (economic) well-being of other inhabitants of the Earth. But he went further than human tyrants in pride and the lust for domination, being in origin an immortal (angelic) spirit. In *The Lord of the Rings* the conflict is not basically about 'freedom', though that is naturally involved. It is about God, and His sole right to divine honour.[41]

To be human, therefore, is, in the moral respect, to be among those who have interior knowledge of their ability to distinguish right from wrong, better from worse, good from evil, and who can act and judge on that basis. *Whether* they do so is precisely the issue of conscience, a word and a concept evoked phenomenologically from the unequivocal recognition within us of moral demand and moral facts: "'I wish I had known all this before,' said Pippin. "I had no notion what I was doing." "Oh yes you had," said Gandalf. "You knew you were behaving foolishly; and you told yourself so, though you did not listen."'[42]

Such a claim about what it means to be human in the moral respect needs of course to be qualified or at least refined. Conducive properties cannot be seen by those who are blind, nor heard by those who are deaf. Nor are people equal in their capacity to act, as Tolkien observed of Gollum:

> There exists the possibility of being placed in positions beyond one's power. In which case (as I believe) salvation from ruin will depend on something apparently unconnected: the general sanctity (and humility and mercy) of the sacrificial person. ... Gollum had had his chance of repentance, and of returning generosity with love; and had fallen off the knife-edge.[43]

The claim, therefore, is not that all moral facts lie 'out there', with a kind of independent objectivity, whether people are here to observe them or not. Conducive properties evoke emotions and rational reflection in the sort of beings that humans happen to be, constituted in the way that they are. Therefore, even though moral and aesthetic statements turn out to be, like other ordinary statements, capable of being true or false in relation to what they purport to be about, there will be no such statements if people are not here to make them. As Keehok Lee makes the point:

Values/norms are indeed human-made products. A world without people is a world without values/norms; a world without people is still a world with objects in it which we now call 'trees' and 'rocks'. A world without people possessing language in the way we do is clearly a world without moral 'ought' propositions. In this sense, values/norms are not a part of the furniture of the world in the way trees, rivers and volcanoes are part of the furniture of the world.[44]

But when we recognize the linkage between conducive properties, emotion and rationality, the claim that 'we make the world' becomes innocuous – or to put it the other way around, it no longer has anything like the subjectivist, non-cognitive consequences that J.L. Mackie supposed when, to his widely influential book, he gave the title, *Ethics: Inventing Right and Wrong*.[45] The word 'invention' seems subjective, but the underlying Latin indicates the reason why it does not have to be so, since the word *invenio* means 'I come into':

So the issue is whether there are *data*, given things, into which we come as we explore and extend the environments in which we live and move and have our being, which do not depend for their existence or nature on our opinion. To say that something is 'invented' is not necessarily to say that it is somehow less than real.[46]

Clearly there are data in relation to moral judgements, because there are conducive properties that initiate the rational and emotional responses with which most humans are familiar. Some there are who refuse or who cannot make the connection, but, as Charles Taylor points out, that does not affect the fact of what normally (norms here being simply statistical) happens:

Moral argument and exploration go on only within a world shaped by our deepest moral responses ... just as natural science supposes that we focus on a world where all our responses have been neutralized. If you want to discriminate more finely what it is about human beings that makes them worthy of respect, you have to call to mind what it is to feel the claim of human suffering or what is repugnant about injustice, or the awe you feel at the fact of human life. No argument can take someone from a neutral stance towards the world, either adopted from the demands of 'science' or fallen into as a consequence of pathology, to insight into moral ontology. But it doesn't follow from this that moral ontology is a pure fiction, as naturalists often assume.[47]

Paradoxically, 'naturalists' in Taylor's sense do not usually espouse the natural normativity which is tied to Aristotelian necessities, and which the direct seeing of conducive properties reinforces. Because there are such necessities, it is easy to see that there such things as moral facts, despite the assurance with which we are told there are no such things. Moral facts are products of the will, or at least of agency, and they are recognized in the consequences of those decisions and actions that become candidates for such judgements. Moral facts lie before us as products of human will and agency, and they can be seen directly. But that does not require that the recognition and the evaluation of them must have the same coercive constancy as the recognition of colour. What is required is that there be reliability and consistency in the community of judgements (as there is), while accepting that this kind of human recognizing and judging is less certain than the recognizing of colour, and is thus more open to contest and change. It is a different kind of judging. But to put it succinctly: if you want to see a moral fact, look at a moral action or even at a moral life, bearing in mind that no life is perfect. It then follows that the recognition of values and norms is not obviously a non-cognitive exercise, still less is it an irrational (because affective) exercise. As Lee puts it:

> Epistemologically, values/norms are part of our cognitive/intellectual operations. We come to know that it is right or justified to adopt certain values/norms/ends by means of our intellectual prowess and faculty. To speak pictorially, it is our intellect (the cognitive side) which persuades our will (the conative side) that it is rational to adopt certain values or to commit ourselves to certain courses of action and not others.[48]

What we now know for the first time is the extent to which emotions are bound into 'our cognitive/intellectual operations' as an ordinary and inextricable part of them.

Building Bridges from Fact to Value: Implication and Entailment

But then, so what? All this will surely remain trivial if it is indeed the case that we can never move from facts to value judgements because there is no logic of implication or entailment that will carry us over from one to the other in such a way that all people must agree in the way they must do in the case of the redness of a book – the third of those three post-Humean, postmodernist claims (pp 76-7): you cannot get values from facts, you can't get an *ought* from an *is*.

But even that, I think, is wrong. It depends very much on what kind of implication is thought to be necessary. If strict and perfect implication is required, then that claim is certainly true. But why are we required to look for strict implication when we do so in hardly any other area of our lives? If strict implication were to be required for all our judgements, we would never get out of bed in the morning: will the floor be there? Will the bread toast? Why should we be asked for more rigorous implication in the moral case than is asked for in most other aspects of our lives?

An immediate answer is that anything less will not deliver an agreement on moral judgement of a two-hands, redness-of-a-book kind. But if we can live with degrees of probability in virtually all other areas of our lives, perhaps it would be a great deal wiser to settle for less stringent degrees of implication here also, the kind of loose implication of which Gilbert Ryle used to remind us when he spoke of the informal logic of ordinary language. On that basis, Keehok Lee has argued for a different degree of implication where moral judgements are concerned, for what is known as epistemic implication.

According to Lee, epistemic implication characterizes the relation between assertion (A) and evidence (E) when (E) is not flippant but serious (thus satisfying referential relevance), when (E) is not only serious but also causally relevant (thus satisfying both referential and causal relevance), when causally relevant and serious evidence (c/rsE) obtains prior to commitment to (A) and is causally independent of that commitment. In those circumstances, where (c/rsE) is true, then (A) would be supported, and where it is false, then (A) would be false. So she concludes:

> Logic textbooks as a rule talk about two kinds of implication, strict implication (entailment) and material implication. But epistemic implication is neither. Strict implication is too strict and material implication is too lax to do the job in hand, namely, to provide rational safe-conduct in the passage from evidence to assertion. ... The burden of this book is to show that although no particular fact uniquely entails a particular 'ought', it does not follow that compatible with the same fact may be conflicting 'oughts'. If epistemic implication holds good, then an 'ought' is only supported if certain facts obtain and not others. It also implies that the support or corroboration is tentative and open to possible future refutation (and the corollary that its lack of support or corroboration is equally tentative, for evidence may turn up in the future to defeat the present verdict).[49]

Epistemic implication thus arises when there is serious and relevant evidence that does entail obligation, but not in any abstract way, only in relation to the circumstances that obtain at the time: your car won't start in the morning: what *ought* you to do? Nothing if you have no appointments that day; ring the AA if you are a doctor on call to a seriously ill patient. There is still entailment but it is contextual and certainly not coercive: it does not entail only one consequence. Nevertheless, it does build a bridge from an *is* to an *ought* in relation to the predication of good in the context in question. It is a human good for doctors to seek the well being of their patients (it is indeed a part of the Hippocratic oath), but it does not coerce them into only one consequence; they might ring up, not the AA, but the RAC, or they might ring up for a taxi, or they might ring one of their partners. As with the Polish bees after Chernobyl, it is context that entails particular behaviours, and context that we are equipped to scan.[50]

So if we bring moral justification more into line with the less severe requirements which we operate elsewhere in life, we can see why it is that non-moral facts do imply obligation, but not with incorrigible coercive-ness. In 2004, scan sequences of a 12-week-old foetus making what appeared to be deliberate and coordinated movements in the womb led to an immediate response on the part of political and health authorities in the UK to investigate whether the upper time limit for abortions should be changed from 24 weeks to at most 18 weeks, or perhaps even 12 weeks. The new facts evoked an emotional *and* rational response, and they im-mediately entailed obligation in relation to the value that human life has. However, this did not coerce the authorities into one single decision or proposal.

To give another example: films or photos of animals undergoing experiments are bearers of conducive properties which evoke reactions and judgements of disapproval in many people. They are certainly emo-tional, but they are also and at the same time rational. In the debate about animal experimentation, philosophers like Tom Regan claim as a matter of fact that some animals are the subject of a life. They 'bring the mystery of a unified psychological presence to the world'.[51] This claim is founded on the observation of causally relevant and serious evidence (to go back to Lee's summary) in the case of animals that exhibit desire and memory, feel emotion and act intentionally. In Kantian terminology, such animals are, like humans, ends in themselves and not means to other ends that may happen to serve humans well. In that case, the facts entail the value that animals have. Insofar as the language of rights derives from value of that kind, then animals have a right to life, liberty and somatic (or bodily) integrity, and certainly they have a right to be treated with respect. If that were not so, then, as Frey has pointed out, there is no argument why we should not experiment on humans as well as on animals.[52]

For some, the facts entail, not only values, but also obligations: the conducive properties entail for them the obligation to act on behalf of such animals, in, for example, the Animal Liberation Front. It hardly needs pointing out that others do not agree. They might, for example, deny the facts, or insist on other facts,[53] or equally they might recognize obligation but might exercise it in different ways. But that in itself does not destroy the logic involved of epistemic implication whereby *oughts* are derived from *ises*, whereby there is a bridge from facts to values.

Disagreement means that the implication is not strict: it is not coercive. Many, for example, might recognize the conducive properties in the films and photos of animals undergoing experiments, but might add to the evaluation of the issue the price that would have to be paid if medical research on animals were to be given up – perhaps by a kind of utilitarianism. Even then, a demonstration of the benefits humans would lose if they gave up animal research does not in any way prove that animals have no moral rights: neither the claim that humans have superior cognitive abilities and are doing this for the benefit of animals as well as of humans, nor the fact that animals often treat each other extremely badly (for example by eating each other) resolves the question of what our moral duties are to animals. For those who make a moral defence of experiments on animals, the only logical step to take would be to adduce other facts that also entail values (for example, the fact of death being a necessary condition of life) and bring that to bear as a countervailing claim.

So cognitivists in relation to moral facts are not defending a position that judgements of good and evil are so certain that all people in the right light and with normal vision must see them with as much certainty as they see the redness of a book; nor do they claim that they cannot be contested and corrected. The cognitive model is not 'knowing that' (like knowing that a book is red by going and taking a look at it) but 'knowing how' (like knowing how to build a dry stone wall successfully, how to look after your patients): facts imply what ought to be done if the outcome is to be successful, and if it is to evoke the judgement that it is good.

Certainly, we will still need to make the decision to live in that way, and that is why there has been an increasing emphasis in recent years on what are known as virtue or character ethics, knowing how to live contextually in ways that evoke the predication of good. The emphasis may be recent, but that way of understanding ethics is very old. It resulted in many lists in different societies and cultures of the conducive properties that lead to judgements of approval: 'He hath shewed thee, O man, what is good; and what doth the Lord require of thee, but to do justly, and to love mercy, and to walk humbly with thy God?';[54] 'the fruit of the Spirit is love, joy, peace, patience, kindness, generosity, faithfulness, gentleness and self-control';[55] faith (*śraddha*), energetic commitment (*virya*), equanimity (*apekṣa*);[56] clarity of mind (*viveka*), freedom from desires (*vimoka*), concentration without distraction (*abhyasa*), attending to one's own responsibilities (*kriya*), speaking the truth (*kalyana*), equanimity in physical distress (*anavasada*), being content with one's own condition (*anuddharśa*).[57] Zizhang asked Confucius about the character of the truly human (ren/jen)[58]: he was told that it can be seen in the five practices, which are respect for others (gong), tolerance (kuan), good faith (i.e. being trustworthy, xin), effort/diligence (min), generosity (hui).[59] All these are recognizable conducive properties which lead to judgements of approval that are shared in particular societies. The lists are different, but the process of conduction from properties to approval is common to human beings, then and now.

In that strong and important sense, there cannot be 'der mann ohne eigenschaften', the man without properties, the man without qualities, to quote the title of Robert Musil's novel. Musil wrote his novel during the years of Hitler's rise to power, and he placed the work of the Parallelaktion in the year 1913-14. In the Committee, the discussion searches for foundations on which the collective life of the people can be built, ideals by which its life can be guided. It focuses on events and people of the time. One of these is Moosbrugger, a psychopath who has murdered and mutilated a prostitute. Should he be held morally responsible and executed? Or should he be regarded as insane and therefore not responsible? In the discussion, the members of the Committee cannot identify any absolute standards for evaluating the act, least of all in the public press, which celebrates in detail and with relish the shock and horror of the act, while denouncing it as evil. They come to the relativist conclusion that they cannot even be sure that it is a crime, and Ulrich, the man without properties, listens to their argument, knowing as he does their own concealments and secrets, and comments, 'If humanity could dream collectively, it would dream Moosbrugger.'

But the argument advanced here shows the reason why 'humanity', even if it did dream that dream, would know, both rationally and emotionally, why to call it a nightmare. In 'the formation of the moral self', to quote the title of a recent book,[60] the conducive properties endorsed in religious traditions have played a fundamental part in producing people who can, appropriately, be called 'good'. The tension for people in those traditions is how to deal with challenges to those properties, when the characterizations of life that they have produced seem no longer to be good. Nietzsche's challenge to Christianity cut very deep, because, as Edith Wyschogrod recognized, it called into question the meaning of the word 'good' in relation to those who had, in one part of the Christian system, been thought of as exemplary, those who have been designated saints.[61] As I put the question briefly in Is God a Virus?, 'In this Nietzschean perspective, who would be a saint?'[62] Yet even if the properties and their embodiment are always under review, it does not alter the way in which conducive properties are directly observed, or the way in which their characterization of lives evokes judgements of approval and disapproval.

Character and Rule Moralities

Thus the development of 'character' or 'virtue' morality, in distinction from 'rule' morality, shows how closely it is linked to the direct seeing of conducive properties, even though those properties are under review, or open to challenge. At the Conservative Party Conference in 1993, John Major issued his appeal for the country to return to basics:

> It is time to return to the old core values. Time to get back to basics: to self-discipline and respect for the law; to consideration for others; to accepting responsibility for yourself and your family – and not shuffling it off on the State. I believe that what this country needs is not less Conservatism, it is more Conservatism.

So what are those basics to which he wished to return? They rest on conducive properties that can be publicly recognized. In an earlier speech, he had looked back nostalgically to the kind of England he hoped to conserve: not just compulsory games at school, but, as he put it, 'Long shadows on county grounds, warm beer, invincible green suburbs, dog-lovers ..., old maids bicycling to Holy Communion through the morning mist.'

Obviously, those particular properties might not lead all people to judgements of approval. But all people can recognize them. Mr Major was not being in the least original in specifying those particular properties, since he was in part quoting (unacknowledged) from someone at the opposite end of the political spectrum, George Orwell:

> The clatter of clogs in the Lancashire mill towns, the to and fro of lorries on the Great North Road, the queues outside the Labour Exchange, the rattle of pint-tellers in the Soho pubs, the old maids biking to Holy Communion through the mists of the autumn morning – all these are not only fragments, but characteristic fragments of the English scene.[63]

Not only are they characteristic fragments, they are *characterizing* fragments: they confer character on particular ways of living, and here again the connections with conducive properties are clear. They are not exhaustive, but equally, they are not private. In 1943, John Betjeman gave a talk on the BBC Home Service, printed in *The Listener* under the title 'Oh to be in England ...':

> I do not believe we are fighting for the privilege of living in a highly developed community of ants. That is what the Nazis want. For me, at any rate, England stands for the Church of England, eccentric incumbents, oil-lit churches,

Women's Institutes, modest village inns, arguments about cow-parsley on the altar, the noise of mowing machines on Saturday afternoons, local newspapers, local auctions, the poetry of Tennyson, Crabbe, Hardy and Matthew Arnold, local talent, local concerts, a visit to the cinema, branch-line trains, light railways, leaning on gates and looking across fields; for you it may stand for something else, equally eccentric to me as I may appear to you, something to do with Wolverhampton or dear old Swindon or wherever you happen to live. But just as important.[64]

There is not a single item here that occurs in the list of either Orwell or Major. That is not surprising. Appeals to conducive properties may list instances in which they are exemplified, but they do not attempt to turn the examples into rules or codes of behaviour – or if they do, as Mr Major rapidly found out, they are likely to sink beneath the waves of public derision. These appeals are doing something different: they are trying to articulate those common assumptions in a society which will act as a premise for the formulation of social and individual actions of a kind that can be called ethical and moral.

To put it more simply, they are listing some (among the no doubt many) constituent properties that characterize 'a good society'. They are what the sociologist Alvin Gouldner used to call 'the background assumptions' and 'the domain assumptions'[65] that characterize a society one way rather than another. They are the often implicit but commonly shared assumptions in a particular society on the basis of which approval and disapproval become possible, not least, if necessary, in the form of legislation. They underlie the intelligibility of that general and unspecific command of every childhood, 'Behave yourself!' Although it is unspecific, all of us knew exactly what it meant. It was a command to come back inside that ring-fence of acceptable behaviour that relies on observable conducive properties to such an extent that it does not need spelling out in detail for it to be immediately understood. While they *may* be expressed in rules or even in laws, they do not have to be so. They may be expressed even more in the character and behaviour of people in themselves and in society.

Character and rule moralities are two different ways in which philosophers have tried to determine what morality is about: is it about the acts and the rules by which people live (or should live), or is it about the formation and expression of character?

The divide, roughly, is between those who shun the relativity of moral judgements and reject the observation that moral facts, if there are any, are not coercive, and those who hold that moral judgements may differ greatly, depending on whether you live in Wolverhampton or 'dear old Swindon'.

For the former, the overriding purpose of ethical and moral philosophy will be to overcome the apparent relativity and subjectivity of all moral and ethical judgements, either by identifying an objective source of morality, or by finding a disinterested and neutral point of view from which to resolve moral conflicts or disagreements. Morality should become an independent guide to the project of life, bringing timeless rationality, truths and concepts to bear on particular issues. If that cannot be done, then we would be left in a world of relative opinions on moral issues, in which moral experience can be a guide, but in which morality has no authority, exactly as Henson feared (p 37). It was basic to Kant's argument that nothing contingent can have universal authority, and if morality does not have independent authority, it cannot function with the force of moral law. It becomes a matter of opinion or of inclination whether we follow its precepts or not. So he stated:

> ... the ground of obligation here must not be sought in the nature of man or in the circumstances in which he is placed, but sought *a priori* solely in the concepts of pure reason. ... Every other precept which rests on principles of mere experience, even a precept which is in certain respects universal, so far as it leans in the least on empirical grounds (perhaps only in regard to the motive involved), may be called a practical rule but never a moral law.[66]

As we have seen (pp 81, 84), the quest for moral objectivity in religions often issues in strong systems of command and control. It is the basic commitment of what Jowitt has called (in relation to politics), 'Joshua discourse', a belief that there is a clear authority, or a transcendental rationality, which is working toward a world 'centrally organised, rigidly bounded, and hysterically concerned with impenetrable boundaries'.[67]

Less aggressively, the aim of ethics, from this point of view, must be to overcome or at least ameliorate the consequences of one person's meat being another person's poison: the aim of ethics will be to set aside the personal judgements and interests of the agents in moral behaviour in order to establish a transcendental judgement, or if not transcendental, then at least an authoritative consensus on what counts as a good or evil act, based on rational considerations.

On the other side in this the debate are those who think that the objective view of morality which issues in act or rule morality, whether it takes the form of utilitarianism at one extreme or of deontology at another, whether, in the religious case, it issues institutionally in Taleban Islam or in encyclical Catholicism, rests on a mistake about ourselves. The truth is that we make our moral choices in particular bodies, histories and societies. We do not learn to be moral beings from books about ethics, still less from attending to moral philosophers. We learn to be moral beings in families that are themselves embedded in particular societies. We may endorse what we have learnt in our subsequent behaviours, or we may deny and contradict it; and here surely is fruitful ground for conflict and dissent on particular issues. But we are unlikely to resolve those issues by trying to lift the argument about them out of the embodied and em-bedded circumstances that have given rise to them, and in which judgements and decisions have to be made. We learn to be moral beings by being them. To adapt Rodin's remark (p 58), character is the essential truth of any person, whether ugly or beautiful, whether evil or good.

Those who take this view then argue that the focus of moral reflection should not be so much or so exclusively on the search for the abstracted grounds of morality, nor on how we can overcome the relativizing truth that we cannot readily or easily gain general consent on the evaluation of particular acts. Far from wishing to get motive and intention out of the way of an independent rationality (as Kant proposed), this way of under-standing morality regards it as far more realistic to recognize that people exist before they act. They therefore bring to their acts (and to their evaluation of them) their own history and biography; and that is why we learn more about the actual living of moral life from novelists and histori-ans than we do from moral philosophers. To read Dickens or Trollope or George Eliot, or for that matter, William Burroughs or Thomas Keneally, is not only to learn about the complexities of moral life; it is also to be affected in the living of our own life by what we read.

That is why the endless debate about the influence of violent videos on young people almost invariably misses the point. The research usually focuses on the difficulty of establishing a link between a particular video that an offender has seen and a particular act of violence. Sometimes a direct influence can be shown, but very rarely, and so the conclusion is usually drawn that violent videos are neutral, since many people watch them who never commit a violent act. But that is to combine an ab-stracted act morality with a failure to understand the difference between cause and constraint. It asks whether we can identify the cause of a par-ticular act (instead of the multiple constraints), and whether we can identify a direct causal link between a violent video and a particular act. A morality that focuses on the formation of character will reflect in a very different way on the relation between what a person sees or reads and the consequence by way of constraint in the formation of character over time.

Thus in 2004, a case came to court involving a particularly savage murder of a 14-year-old by an older boy. It was claimed by the parents of the victim to be a direct consequence of the older boy's obsession with the video game 'Manhunt', in which players are rewarded for inflicting the most brutal deaths they can. The game was immediately withdrawn from sale by some retailers, but the director-general of ELSPA (the Entertainment and Leisure Software Publishers Association) responded: 'There is no substantive evidence in this case to link this tragic event to the fact he was playing a game. The police have confirmed that they found Manhunt in his room, but there was no mention of it during the court case.' The point here is that a direct influence can sometimes be shown, but very rarely, but that in any case ignores the effect on the formation of attitudes and character.[68]

So this other way of approaching morality, a morality of character or of virtue, starts from the fact that the person precedes the act. To be a person of virtue, or of a character that can be appropriately described as 'good', is to have learnt the exercise of moral skills in a particular context, and to have learnt also the extent to which they are reliable in a wider world.

It is true, of course, that a person's character is formed in a particular body, history and society, which will certainly have rules and laws (and the attitudes that people take to rules and laws is very much a part of the formation and exhibition of their characters). Those histories and societies vary greatly, and that means (as we have already seen, pp 96–7) that there will be many different notions in the world of what counts as a virtuous or admirable character. There may well be convergence among these notions at many points, but even so, there will be no single list of virtues that are agreed to constitute a virtuous character.

Nevertheless, what *is* universal is the human ability to make evaluative judgements of character on the basis of conducive properties, and that is not affected by the fact that we may be mistaken. The consequence of that direct seeing of conducive properties was put well by Samuel Johnson, in *The Lives of the English Poets*:

> One man may know another half his life without being able to estimate his skill in hydrostatics or astronomy; but his moral or prudential character immediately appears.

Chesterton made the same point when he observed that the goodness of a person is something that can be directly encountered and known, as realistically as a pain or a smell. When Mr Blair made, as prime minister, his reply to the Queen on the occasion of her Golden Jubilee, he congratulated her, not for a list of particular achievements (i.e. conducive properties), though doubtless he could have done so, but for what he called a recognizable quality: 'It is a quality that is the very best of the British character and when we find it, we recognise it immediately.'[69]

Chesterton's way of putting it was overstating the case, because the encounter with goodness is nothing like so incorrigible as he claimed. We are far more often deceived than he allowed, not least by hypocrites, and it may be all too frequently the case that, as Tallulah Bankhead famously remarked of a play that she was watching, 'There is less in this than meets the eye.' Nevertheless, conducive properties are seen directly, and they often entail obligations and judgements. When John Betjeman said of an archbishop that he was so hard he would have boiled his potatoes in a widow's tears, many would have strongly contested that judgement, even though Roald Dahl's autobiography goes a long way to confirm it.[70]

Does it follow that we have to choose between these two ways of describing and evaluating morality and ethics, between rule moralities and virtue or character moralities? Clearly not, because to do so would be yet another example of the fallacy of the falsely dichotomous question (pp 77–8). Both ways of understanding what is involved in making ethical and moral judgements are needed, and neither can do the whole work of ethics on its own. But in *both* cases, the direct seeing of conducive properties is fundamental and basic in forming moral judgements.

On their own they are far from being enough. They are a means toward an end that lies beyond them, the practice of a moral life that respects rules (in a critical way) and lives in a characterized way which deserves the predication of good. Ethics is not so much a matter of command and consent as it is of demand and intent. Moral vision (to quote the title of David McNaughton's admirable introduction to ethics[71]) enters into ethics in every possible way, just as the vision and purpose of an artist enters into the work that she or he produces – either as one particular work or as the quest of a whole working life (pp 61, 66–7). As Stanley Spencer, looking back on his life as an artist, wrote of his early paintings:

> Before the 1914 war I had to be very convinced about a picture before I drew it or painted it. The drawing or painting of the thing was the experiencing of heaven: it would have been unthinkable that I should or might find hitches or snags. ... If I know that a certain result will arise from a certain experience I value the experience and I feel quite sure long before I have seen the result. For instance I believe in my associations and always have done. I believe in them as having the power of realising the meaning I seek because they are guided by another and deeper belief, namely that only goodness and love and Christian and other benign beliefs are capable of creative works.[72]

When Spencer used the word 'only' in that last sentence, he was clearly writing about himself, not about, for example, Goya. 'Enduring creation' (to quote the title of Nigel Spivey's book) and envisioning evil have been creative for many artists, especially 'after Auschwitz'. In a parallel to the fundamental question whether theology is possible 'after Auschwitz', Spivey asks whether art 'after Auschwitz' is possible. Of his first chapter, 'Stepping Worstwards', he wrote:

> The opening chapter is an overture: partly confessional, broadly moralizing, and, I should say, obligatory, since there is a strong case for arguing that 'the Western cultural tradition' went to ground at Auschwitz, and was thoroughly mocked by what happened, at Europe's centre, under National Socialism. I deny the notion of anyone being 'a Holocaust bore'; yet I admit that I write as simply one more belated witness.[73]

The reality of evil means that moral vision necessarily includes what Michael Berenbaum called, 'the vision of the void'.[74] It means also that the contest against evil is unending and constant. Henson cleared the books from the bookcase opposite the desk in his study, and placed there, constantly before his eyes, a crucifix. Others might place an image of a bodhisattva or of Rama or of one of the many other ways in which, in order to enable resistance, God enters into ethics, not least as the vision and goal of all goodness in a way that acts as a constraint over behaviours in the present – and where that goal lies in the future, this is an instance of what is known in biology as 'downward causation'.[75]

In the work of art that is a human life, the struggle is constant, to bridge the gap between the vision and the reality of what we are, and it is a major part of all religions to do exactly that – 'to close the gap', as Peter Byrne puts it, 'between value and reality, between the world as it ideally should be given human values and the world as we find it',[76] given also the conducive properties that engage us and the choices (including refusals to do anything) that lie before us. The decisive intervention of religions into ethics lies in their insistence that humans cannot do this effectively on their own, and that they therefore need the grace of God to help them (though religions have different words and different understandings of what exactly that means). On that basis, religions demand or at least promote this way of being moral, in which intention, act and value are connected, because they set the whole enterprise of ethics in the context of value which is believed to transcend the contingent. Transcendental value is required as a foundation for moral persistence and moral hope. In a brilliant book, recording how he repeated Thoreau's 'retreat to Walden Pond', Erazim Kohak wrote:

What is crucial is that humans, whether they do so or not, are capable of encountering a moment, not simply as a transition between a before and an after, but as the miracle of eternity ingressing into time. That, rather than the ability to fashion tools, stands out as the distinctive human calling. Were it not for humans who are able to see it, to grieve for it, and to cherish it, the goodness, beauty and truth of creation would remain wholly absorbed in the passage of time and pass with it. It is our calling to inscribe it into eternity.[77]

Thus God enters into ethics, not as a source of control alone but as the source of value in the creation of beauty from the beast, and, even more to the point in practice, as a real presence in the encounters of life. Those encounters are collective as well as individual, and it means that ethics has as much to do with public as with private excellence. As George Woods argued:

> The quest for moral perfection is far more than a quest for private excellence. It involves the attainment of right relationships with God and with our fellow-men. And it includes a right relation of re-created humanity to the natural world. The consummation is set in eternity, which lies far beyond the range of profitable description.[78]

The importance of conducive properties is that they occur, not in just in eternity, but now, and that they help us to see moral facts directly. That is not affected by the truth that our judgements are corrigible: being able to acknowledge that I was mistaken and to say that I am sorry is a part of the formation of a morally virtuous character. Indeed, the seriousness with which we take ourselves as moral beings will be measured by the extent to which we recognize moral failure in our own case, and do something about it and about its consequences. Aristotle quoted Agathon as saying, 'Even God cannot change the past.' But in what I have called elsewhere 'retrogressive rituals', that is exactly what God does do. What happens in ritual is that the changing of the past becomes objectively a fact, a new fact, because through ritual 'the past' becomes *our* past in a new way.

Reason and Emotion

In the forming of a moral life, conducive properties are fundamental. They are not coercive properties, nor are they acting as a substitute for a 'property of goodness'. On the other hand, they can act as a means of grace, since they are independent of our own construction, and they offer important reasons why people arrive at the judgements they do. They therefore open the way to a rational discussion of them.

But that is only possible because we are, all of us, built in such a way that we do observe conducive properties so consistently that they evoke both moral and aesthetic judgements in the sort of beings that we happen to be, constituted in the way that we are. We may construct from them as many diverse moral systems as there are styles of art, but there remains in common the constitution that enables us to do this at all.

It is a constitution very different from what Hume took to be 'the fabric and constitution of the human species'. Does the difference matter? It clearly matters greatly if we are to translate the new integration of emotion and reason into the realities of human life, whether in the policies of a society (for example, in education) or in the forming of character in individual lives.

The old view took for granted that our feelings and emotions are sharply distinguished from rational reflection and control. What, in contrast, the recent work on the neurophysiology of emotion, feelings and reason makes clear is that the human experience of value does not always and invariably come to us in two parts, an emotional perception of what is there to be observed, followed by a rational evaluation of what has been observed.

Even where that *does* seem most obviously to be the case, it may not actually be so. For example, virtually everyone knows what P.G. Wodehouse was describing when he wrote, in *Uncle Fred in the Springtime*, 'Pongo spoke a little huskily, for he had once more fallen in love at first sight.' Conducive properties lead to a direct 'turning on', in which emotions are engaged, and it has immediate consequences through the whole body. Wodehouse also wrote: 'What magic there is in a girl's smile. It is the raisin which, dropped in the yeast of male complacency, induces fermentation.'[79]

But even in that most obvious case, it is by no means *always* the case that a rational scan or evaluation is added on only at a later stage. We are not always and simply 'swept off our feet'. For what we now know is that the neurophysiological processes involved connect our direct seeing of conducive properties to the whole somatic process, including rational scanning, in a way that binds the whole process to biography – to memory and to the personal and socialized history that each of us has (or maybe one should say, that each of us is) – hence again the importance of character or virtue moralities.

This means that when Hume wrote that there is nothing, in itself, valuable or despicable, desirable or hateful, beautiful or deformed, but that these attributes arise from the particular constitution and fabric of human sentiment and affection, he was almost entirely right but actually missed the target. Yes, our perceptions, emotions, feelings and rational scannings do arise from the particular constitution and fabric that underlie human sentiment and affection: from where else could they possibly arise? Infants who are only two to three weeks old can already imitate facial movements, including sticking out the tongue, opening the mouth and pushing out the lips. Since this happens long before a new-born baby can start looking in a mirror, we have to infer some kind of interior map that enables an infant to know which of its own facial muscles correspond to those of another human being.[80] They then also associate facial expressions with corresponding attitudes, so that, as the phenomenologist Max Scheler pointed out, they can be directly aware of playground attitudes on their first day at school without necessarily having had to learn to decode them previously in the family.

But far beyond infancy, the major gain of all this recent work is that it binds together the emotions and rationality in an indissoluble way, and it connects them with the factual limits set by the worlds in which we live. Our understanding of the constitution and fabric of our humanity is changed completely from the division between the two that Hume took for granted in one way, and his opponents in another (p 40). We are not the sort of split creature that has been so often described in the past – by Milton, for example, when he wrote of the Fall as the surrender of Reason to rebellious and wayward passions. In *Paradise Lost* he wrote:

> Reason in Man obscur'd, or not obey'd,
> Immediately inordinate desires
> And upstart passions catch the government
> From Reason, and to servitude reduce
> Man, till then free.[81]

This radical separation of reason and emotion does not belong only to the distant past. It was reinforced during the twentieth century when the emotions were thought to be a vestigial survival from our early evolution. On this view (and it was a commonplace among brain and behavioural scientists), the limbic system is the resource of the four 'Fs' of survival (feeding, fearing, fleeing and sexual reproduction), while the neocortex controls the basic system in a cool and rational way. As Damasio describes the old picture:

> In simple terms: The old brain core handles basic biological regulation down in the basement, while up above the neocortex deliberates with wisdom and subtlety. Upstairs in the cortex there is reason and willpower, while downstairs in the subcortex there is emotion and all that weak, fleshy stuff.[82]

On that picture, it was held to be a mark of the truly human that people controlled their emotions and brought them under rational command. Why else, having already gone blind, write an epic poem of many thousands of lines? The answer at that time was 'education, education,

education'. In Dryden's elliptical comment: 'The design of it [the heroic poem] is to form the mind to heroic virtue by example; 'tis conveyed in verse, that it may delight, while it instructs.'[83]

In that spirit, Milton wrote that his work was to be 'doctrinal and exemplary to a nation'. He claimed that the work of any poet should be:

> ... to inbreed and cherish in a great people the seeds of vertu, and publick civility, to allay the perturbations of the mind, and set the affections in right tune, to celebrate in glorious and lofty hymns the throne and equipage of Gods Almightinesse, and what he works. ... Teaching over the whole book of sanctity and vertu through all the instances of example with such delight to those especially of soft and delicious temper who will not so much as look upon Truth herselfe, unless they see her elegantly drest, that whereas the paths of honesty and good life appear now rugged and difficult, though they be indeed easy and pleasant, they would then appeare to all men both easy and pleasant though they were rugged and difficult indeed. And what a benefit this would be to our youth and gentry ... .[84]

What a benefit it would be, in other words, to those most prone to be swept away by passion and emotion. On that view, emotions and feelings are the enemy within – 'Ira furor brevis est': 'Anger is a brief fit of madness'.[85] That well-known saying underlies a legal defence in the case of crimes committed under emotional duress, a *crime passionnel*, as the French would say, a crime committed when one is swept away by emotion and has lost rational control. In the more colourful sentence of Tristram Shandy, 'When a man gives himself up to the government of a ruling passion – or in other words, when his Hobby-horse grows headstrong, – farewell cool reason and fair discretion!'

If that is so, then it should be a fundamental part of education to learn how to bring and how to keep one's emotions under control; and that is exactly the point that Horace was making in that section of the poem from which the phrase, *ira furor brevis est*, actually comes:

Qui non moderabitur irae,
Infectum volet esse, dolor quod suaserit et mens
Dum poenas odio per vim festinat inulto.
Ira furor brevis est: animum rege; qui, nisi paret,
Imperat: hunc frenis, hunc tu compesce catena.
Fingit equum tenera docilem cervice magister
Ire viam, qua monstret eques. Venaticus, ex quo
Tempore cervinam pellem latravit in aula,
Militat in sylvis catulus. Nunc adbibe puro
Pectore verba puer; nunc te melioribus offer.
Quo semel est imbuta recens, servabit odorem

Testa diu.[86]

The disastrous consequences of that in British education are familiar indeed. To take an entirely random example: searching for an article I had once read on 'how to keep crocodiles in your garden', I was looking through *The Boy's Own Annual* for 1898. In it there is a remarkable article on the archbishop of Canterbury as a footballer. The archbishop, who went to Blundells at the age of 12, recalled in a speech at an Old Blundellians dinner, how, when he first went to the school, 'he had an exceedingly passionate temper'. 'But,' he continued, 'it was a good thing to have one's temper knocked out of one early in life, and he owed that advantage to Blundells.'[87] It is Lady Circumference asking a master at her son's prep school how her son is doing. On being told 'Quite well', she replies 'Nonsense! ... The boy's a dunderhead. If he wasn't he wouldn't be here. He wants beatin' and hittin' and knockin' about generally, and [even] then he'll be no good.'[88]

Of course there are moments in every life when it is important to control one's emotions, and it belongs to the rationality of emotions, as de Sousa argued in his book of that title,[89] to monitor the appropriateness of the targets towards which our emotions are directed. Equally, it is surely healthy to express one's emotions on many occasions and not always to try to suppress them.

But that way of talking continues the mistake of supposing that there are two sources of behaviour, the rational and the emotional, whose coexistence in a single subject is a matter of selection, as though now one, now the other, ought to be in the ascendancy. On the old model, it is as though I have two competing horses in a single harness, so that unless I can control them and bring them into balance, the carriage of my life will never be drawn safely forward.

The two horses are often called 'the head' and 'the heart', and many of the old accounts of reason and emotion, while recognizing the two, give priority to one over the other. For Aquinas, it is reason or intellect that should govern the body as a despotic power would govern a slave, and should 'control the irascible and concupiscent appetites by presiding over them as a quasi-political power'.[90] On the other hand, Hume (as we have seen, p 40) turned that argument inside out by claiming that reason cannot control the emotions, and that the passions lead the way: 'Reason is, and ought only to be the slave of the passions, and can never pretend to any other office but to serve and obey them.'[91] For Pascal, famously, 'The heart has its reasons which reason does not know', and therefore 'It is the heart which experiences God, and not reason: this, then, is faith,

God felt by the heart, not by reason.'[92] Scheler, in his ambitious project of a phenomenological ethics, followed Pascal closely in this respect:

> The heart possesses, within its own realm, a strict analogue to logic, which it does not, however, borrow from the logic of the intellect. As the ancient doctrine of *Nomos Agraphos* [the unwritten law] can already teach us, there are laws written into the heart which correspond to the plan according to which the world, as a world of values, is built up. It can love and hate blindly or with evidence, just as we can judge blindly or with evidence.[93]

In contrast, the neo-Thomist philosopher Jacques Maritain applied the emphasis of Aquinas on connatural knowledge, the harnessing of the two together in ways that cannot be disentangled:

> It is through connaturality that moral consciousness attains a kind of knowing – inexpressible in words and notions – of the deepest dispositions – longings, fears, hopes or despairs, primeval loves and options – involved in the night of the subjectivity. When a man makes a free decision, he takes into account, not only all that he possesses of moral science and factual information, and which is manifested to him in concepts and notions, but also all the secret elements of evaluation which depend on what he is, and which are known to him through inclination, through his own actual propensities and his own virtues, if he has any.[94]

'Connaturality' was, in its time, anticipating the way in which the separation of emotions and rationality has now been called in question. As John Ratey has put it:

> The new view shows that emotion is not the conveniently isolated brain function that we once were taught. Emotion is messy, complicated, primitive, and undefined because it's all over the place, intertwined with cognition and physiology.[95]

This means in effect that the emotions are related to reason in an already integrated way, and that emotion 'is played out under the control of both subcortical and neocortical structures'.[96] Damasio summarized in five connected points his argument about the common biological core that underlies the integration of emotions:

> 1. Emotions are complicated collections of chemical and neural responses, forming a pattern; all emotions have some kind of regulatory role to play, leading in one way or another to the creation of circumstances advantageous to the organism exhibiting the phenomenon; emotions are *about* the life of an organism, its body to be precise, and their role is to assist the organism in maintaining life. 2. Notwithstanding the reality that learning and culture alter the expression of emotions and give emotions new meanings, emotions are biologically determined processes, depending on innately set brain devices,

laid down by a long evolutionary history. 3. The devices which produce emotions occupy a fairly restricted ensemble of subcortical regions, beginning at the level of the brain stem and moving up to the higher brain; the devices are part of a set of structures that both regulate and represent body states. ... 4. All the devices can be engaged automatically, without conscious deliberation; the considerable amount of individual variation and the fact that culture plays a role in shaping some inducers does not deny the fundamental stereotypicity, automaticity, and regulatory purpose of the emotions. 5. All emotions use the body as their theater (internal milieu, visceral, vestibular and musculoskeletal systems), but emotions also affect the mode of operation of numerous brain circuits: the variety of the emotional responses is responsible for profound changes in both the body landscape and the brain landscape. The collection of these changes constitutes the substrate for the neural patterns which eventually become feelings of emotion.[97]

However, emotions are *about* much more than the body. It is because emotions and feelings are integrated with rational cognition that emotions are themselves genuinely cognitive of much more than whatever will 'assist the organism in maintaining life'. It is the 'much more' that is involved in the direct seeing of conducive properties that leads to ethical and aesthetic responses. Damasio, in contrast, argues that emotional cognition is cognition of somatic (bodily) states:

> *Feelings are just as cognitive as any other perceptual image*, and just as dependent on cerebral-cortex processing as any other image. To be sure, feelings are about something different. But what makes them different is that they are first and foremost about the body, that they offer us *the cognition of our visceral and musculoskeletal state* as it becomes affected by preorganized mechanisms and by the cognitive structures we have developed under their influence. Feelings let us *mind the body*, attentively, as during an emotional state, or faintly, as during a background state. They let us mind the body 'live', when they give us perceptual images of the body, or 'by rebroadcast', when they give us recalled images of the body state appropriate to certain circumstances, in 'as if' feelings.[98]

This 'somatic marker' hypothesis[99] has been criticized by Rolls[100] on the grounds that it requires specific activity in the somatosensory cortex or some other brain region, since without this 'the somatic marker hypothesis would vanish'; however, it seems unlikely that an emotion produced by reinforcement in the body 'would *require* activity in the somatosensory cortex to feel emotional or to elicit emotional decisions'. It has been criticized even more severely by Bennett and Hacker on grounds of conceptual confusion.[101]

Those criticisms simply underline the ways in which emotions can be genuinely cognitive in a much wider and more ordinary sense. That is why it is true to say that in general emotions have objects. As de Sousa argued (p 109), they connect us to the world of experience in ways that may be life-saving, but in many other ways as well. For Rolls, that is an implication of his own definition of emotions:

> The definition given provides great opportunity for cognitive processing (whether conscious or not) in emotions, for cognitive processes will very often be required to determine whether an environmental stimulus or event is reinforcing. Normally an emotion consists of this cognitive processing which results in a decoded signal that the environmental event is reinforcing, together with the mood state produced as a result. If the mood state is produced in the absence of the external sensory input and the cognitive decoding (for example by direct electrical stimulation of the amygdala ...), then this is described only as a mood state, and is different from an emotion in that there is no object in the environment towards which the mood state is directed.[102]

Among those 'objects in the environment' are those which are the bearers of conducive properties. They evoke the characteristic ways in which humans experience their environment and their own interior nature, in which body, reason and emotions are integrated. That is exactly what happens in the human experience of both beauty and goodness, because conducive properties invade the whole system, not just parts of it (Housman's hair bristling while he shaved when his memory evoked emotion and brought into consciousness a powerful line of poetry, p 69). It is because this is so that people who have suffered neurological damage in specific sites of their brains show a consistent 'before-and-after' change in behaviour: for example, in the case of specific damage, some people can call up knowledge of the world around them and can deal with it logically, but their decisions become irrational (damaging to themselves and others) because they are no longer informed by 'signals hailing from the neural machinery that underlies emotion'.[103]

Where, however, the integration of reason and emotion has not suffered impediment because of physical damage, the consequence is the amazing array of characteristic human behaviours. Some of these may well be shared with other animals. To give an example, those who study animal behaviour (ethologists) often speak of 'ritual displays' in the approach of animals to mating, or in such things as the defence of territory. Rituals are, nevertheless, in the human case, entirely different.[104] The difference is the result of this integration of emotion and reason, which humans pursue and cultivate. It is a point argued strongly by W. John Smith, concluding his paper on 'Ritual and the Ethology of Communicating':

This proposal that human ritual is not a fundamentally new class of communicative behavior peculiar to our species does not imply that it has no human peculiarities. ... The enormous cultural elaboration of our rituals correlates with their extension into peculiarly human enterprises such as worship and government. Human beings have reflected upon rituals, their causes, and their uses and have modified them to adjust their functioning. It would be fallacious to assume that human rituals are nothing more than elaborate versions of nonhuman ones, simply because a common origin is claimed. Human social behavior, managed with the highly specialized tool of language, is much more intricate than that of any other species. From studying biological origins and from comparing diverse species, we can gain insights into the properties of such classes of behavior as displays and formalized interactions. But to discover what human beings can do with such classes and how human activities are constrained by them, we must study human behavior.[105]

Ritual and Conducive Properties

That example, of ritual, is particularly important, because rituals make use of all five senses (in the case of Asian religions, all six) in the scan and recognition of conducive properties. They are also commonly reinforced by symbols, because symbols (amongst much else that they do) condense conducive properties and extend them through time. Furthermore, rituals are as common in secular contexts as they are in religious, illustrating the way in which the common human competence to see conducive properties can be used in many different ways.

Rituals are repeated patterns of action and behaviour undertaken for an immense number of purposes – for example, to celebrate the birth of a child, or to lament the passing of an elder; to give thanks for the life-giving presence of food and water, or to bring a death-dealing disaster on enemies; to express praise or penitence; and thus certainly to recognize and come into the presence of God.

For this to work, rituals have to be recognized and understood at the deepest levels of human understanding, including kinds of emotional learning which are going on far below the level of conscious effort or awareness. Rituals link together associative learning and symbolic cognition, two major ways in which human brains process information as they respond to the world. Rituals initiate an interaction between associative and symbolic learning processes by manipulating the sensory characteristics of symbolic displays in such a manner that they are experienced and felt emotionally in very powerful ways: powerful, because reason and emotion work together in ritual in a genuinely cognitive process.

The result is that we live immersed in the rational world of symbols, but we live immersed also *and at the same time* in the emotional experience of symbols. Thus rituals lock reason and emotion together at the most important moments of our lives, and they do it in ways that we can rely on, because rituals are repeated in predictable and well-tested ways over and over again. They are, in other words, highly redundant.[106]

Rituals are thus a fundamental part of building a world in which we can live. We use signs (such things as words, actions, costumes, decorations, gestures) to construct public representations of the world as we happen to understand it. Rituals then attach to these representations an appropriate emotional tone (mystery, awe, solemnity, authority, fear, grief, sadness, joy, exaltation, and so on). Which emotional tone it is depends on the occasion - whether, for example, it is a wedding or a funeral.

All this means that rituals are deliberately created, whether by religious or by secular authorities. They are created in order to carry the conducive properties that will evoke the appropriate emotions: of pride and patriotism at a military parade, for example, of grief at a funeral, or of joy at a wedding - or of reverence in the case of worship.

The relevant conducive properties are therefore brought into being as deliberately in the case of ritual as they are by composers of music or by film producers - or, for that matter, by advertisers: a successful advertisement is one in which the chosen conducive properties strike home.

So it is in ritual. The ritual score is composed as carefully as a musical score, and choreographed as carefully as a ballet. The result, in a successful ritual, is that the appropriate tone or colour is brought to a person (of grief, for example, or of joy), in whom understanding of what is going on and the emotional response to what is going on are simply indissoluble. They are what the person is at that time.

Thus although it may often seem that in ritual emotion overwhelms reason, it is not usually the case. The very power of ritual in combining reason and emotion means that ritual is invaluable in political strategies of persuasion - or indeed, of control. For example, in a book called *The Politics of Reproductive Ritual*,[107] Paige and Paige showed, as anthropologists, how men in small-scale societies use the rituals surrounding human reproduction to keep control over the lives of women. Another anthropologist, Maurice Bloch, argued (in a book called *Prey into Hunter: The Politics of Religious Experience*[108]) for a causative relation between the ritual denial of transience through sacrifice, and what he calls 'rebounding', and often political, violence. In a more general way, Ariel Glucklich has shown how important the ritual use of pain is in mediating 'the flow of power between those who have it and those who do not', especially in rites of passage and of initiation.[109]

But we, surely, are more familiar with this on the much larger scale of totalitarian societies:

Mothers, your cradles
Are like a sleeping army
Waiting for victory –
They will never more be empty.[110]

That is the verse of a song used in the ritual introduced by the Nazis on 10 May each year. It was the day they called the Day of the German Mother, the day when the Mother's Cross was awarded to those 'rich in children' – provided of course that none of the children was in any way dysfunctional, and provided that the mother herself did not smoke or drink, did not sleep around and ran an orderly home. When that sleeping army awoke and joined the Hitler Youth, we know very well the rituals that awaited them, on such occasions as the Nuremberg rallies: ritual as politics by other means.

Even in the slightly more innocent world of British parliamentary democracy, rituals are important, especially in the State Opening of Parliament and the Queen's Speech. Or *were* important: the attitude to ritual of New Labour under Mr Blair was made obvious in the peremptory abolition of the office of the Lord Chancellor. Mr Blair derided the protesting leader of the opposition for defending a man wearing knee-breeches and a pair of women's stockings.

Certainly the rituals surrounding the State Opening of Parliament are far removed from everyday life in London. Richard Crossman in his diary likened the State Opening of 1967 to 'the *Prisoner of Zenda* but not nearly as smart or well-done as it would be in Hollywood. It's more what a real Ruritania would look like – far more comic, more untidy, more homely, less grand.'[111] So ridicule is easy, but what New Labour does not even begin to understand is the way in which ritual reinforces the tenacity of legality. The rituals of a legal system – let us say, of a court room – distance the defence of law and legality from the private opinions of those taking part, or from the manipulations of political control.

Ritual therefore is an example (and only one among many) of the ways in which the human competence to see conducive properties is made use of in human societies. This competence can be relied on in producing ritual, just as it can in producing art or in making moral judgements. Emotions and reasons are equally at work in cognition. Thus cognitivists of this kind do not in the least defend a position that judgements of good and evil, or of beauty and ugliness, are so certain that all people in the right light and with normal vision must see them with as much certainty as they see the redness of a book; nor do they claim that those judgements cannot be contested and corrected.

On the other hand, they do claim that in brains and bodies of the kind that we happen to have, we do see conducive properties which evoke feelings of satisfaction and of dissatisfaction, and which sustain the judgements and vocabularies of moral and aesthetic value.

A consequence of this is that facts do imply values and do imply actions, though *not* in a coercive manner. Conducive properties are not coercive properties. Not everyone will agree with the facts or with the values in relation to human or other good; and of those who do agree, not everyone will go on to agree on how to realize the value in any particular case.

Nevertheless, it is a rational and cognitive exercise, resting on a rational and emotional response to facts, which then attempts to form the will for action. The implication is not strict, but it *is* implication, which does connect facts with values. That the implication is at best probable and rarely certain simply induces in us (or should induce in us) a realization that we cannot always be right, and that we rely, if we are wise, on grace to precede us and mercy to follow us.

If, then, we return to Greville's diary, and to his conventional facts (p 71), we can see that neither in morals nor in history nor in aesthetics can we hope for anything more, certainly not what Greville called 'absolute truth'. But that does not lead to the relativism or to the subjectivity that is so often taken for granted in a postmodernist world. The search for his 'more tolerable and probable case' is not a matter of one opinion set against another. It is consensual because the senses with which we interact with the world and with each other are constructed in each one of us in the same way, a way that is consistently both rational and emotional, and in which limits are set by facts.

Thus a cognitive or critical realist is arguing here, first, that conducive properties leading consistently to moral and aesthetic judgements can be perceived and recognized directly; secondly, that because of the neurophysiological link between emotions and reason, the recognition is not a matter of emotion or sentiment followed separately by rational reflection, still less is it a matter of adding moral sentiment on to the observation of non-moral facts; thirdly, that in any case some non-moral facts do imply moral judgements and actions, and that the is–ought, fact–value, distinction is not as insuperable as two centuries of philosophy have usually taken it to be; and fourthly that, in consequence, the predication of good by constant analogy from one instance to another can be made with confidence on many occasions, as indeed it is.

There is, therefore, a court of appeal in human conflicts, although not the one that Henson envisaged. But is it of any consequence? We are still left with the issue with which we began (pp 2–5): the conflicts between religions which threaten to destroy human life as we now know it. It is to that issue that we will now turn.

5 THE APPEAL TO COHERENCE

At the beginning of Chapter 3 we joined a bishop standing in a dark corridor, contemplating the portraits of his predecessors. Now we join another bishop in the year 1928, and we join him lying, as he put it, on his tummy in a draughty hut in the Holy Land, writing a letter to his wife.

We are doing this because it is necessary now to return to that simmering issue with which we began: the link between religions and both war and terrorism. Those certainly involve emotion. Can reason enter into conflicts as savage and often long-running as those, once we realize that emotions and rationality are not alternatives but are at work together? Clearly we cannot appeal to history as though it contains objective and independent facts that will decide the issue (whatever the issue is in any particular conflict), or as though we can disentangle the historically certain (what Greville called in his diary 'absolute truth') from human interpretations which are a matter of opinion and conflict (Greville's 'conventional facts').

On the other hand, it is now clear that we do make judgements, not just in science, but also in history, aesthetics and ethics, which are far from being subjective and relativistic in the extreme sense (that one opinion has as much or as little warrant as any other). Warrants for our assertions are possible because our judgements are constrained by facts which are independent of our opinions and which we see in consistent and shareable ways.

That is so because our brains and bodies are built in such a way that we perceive conducive properties in objects and people external to ourselves. These conducive properties evoke and sustain the judgements and vocabularies of satisfaction and dissatisfaction (including the satisfaction of truth), and they do so with immense stability and consistency among humans even of different periods and cultures. Thus, to give an example, humans may have different experiences and understandings of time (pp 31–2), but there are nevertheless conducive properties that give them *some* sense of time. It is clear that humans, along with some animals, have a competence to distinguish between past, present and future states.

The human competence goes much further than that of animals in the way that we can project our lives in detail toward those future states – it is a basic reason why there is morality at all, since we cannot know the detail of those futures and have therefore to evaluate our projected actions to decide whether they are wise or foolish, good or evil. Human competence in this area also goes much further than that of animals in the way that memory operates on the past, allowing us even to have a sense of short temporal sequences in the past; there is much about the past that we can retrieve and scan for ourselves as though it is once again in the present. We can indeed write history, even though, as observed, it may always be a present event.

All that seems trivial and obvious, but it is only trivial and obvious because we are so used to doing it, and because almost all people have this time-parsing competence. In fact, it is not yet at all obvious how the brain does this. We can only postulate the competence because we exercise it, and because also we can observe the difference that is made in those who do not have it. Jill Boucher, for example, has done important work on the time-parsing deficit that often occurs in autistic people, in which she starts from the observation of Friedman that 'there is no environmental time that can excite a time sense, only the physical events that that we consider as contained in time'.[1] She concludes:

> This suggests that preconceptual infants, and also animals which, like people, live in a temporal world, must have biologically based mechanisms for analys-ing those temporal aspects of experience which are important for their survival, much as they have biologically based mechanisms for analysing the spatial aspects of experience. It is these primitive biopsychological mechanisms which may be lacking or dysfunctional in some way or other in people with autism, and which have to be effortfully compensated for by those caring for people with autism, and by able autistic people themselves.[2]

Conclusions of this kind are only possible because of the fact that we are built in the same way to interact with the world and with each other. That 'same way' includes the further fact that we are built to do this in ways that integrate rationality and emotion as a kind of reinforcement of each other.

However, we also saw that this cannot ever lead to the *same* judgement of all people in every circumstance. Exactly the opposite: we construct from our responses the brilliant diversity of human culture. But we con-struct also the bitter conflicts in which religions are so often engaged, and which we find so hard to handle. Henson hoped to make an appeal to history in order to resolve those conflicts. But an appeal to history seems more often to exhibit conflict, above all in Christian history where the themes of diversity and contest have issued in the past, as they still do in the present, in such violence and malice that the Christian apologetic in which Henson was also interested is immediately contradicted.

Does that matter? It certainly did to Henson who believed that the case for Christianity (Christian apologetic) can only be made if these conflicts are either resolved or else shown to be – to use the distinguished old word – adiaphoristic.

Adiaphora comes from a Greek word *adiaforos*, meaning in Stoic philosophy 'things indifferent', so that adiaphorism is 'the view that certain items in a controversy are not sufficiently central to warrant continuing division or dispute'.[3]

But how can the more serious conflicts that are not adiaphoristic be resolved? In *Is God a Virus?*, I offered eight 'Guidelines in Conflict'.[4] To them I would now add the argument in this book, *not* as a panacea, but as a reinforcement. It is liberating to understand how conducive properties evoke and sustain the judgements and the vocabularies of value in human discourse with degrees of *corrigible* objectivity, because that brings into more neutral discussion the reasons why particular views are held with such passion. It offers a way of bringing into the open the ways in which emotion and reason can work together.

But even then, does that help in any way to resolve the conflicts which make religions so dangerous and which make the Christian gospel of love seem so ridiculous, since Christians in their institutions are a living contradiction of it? Clearly not on its own. We still have to try to understand, first, why conflicts between religions and among Christians are so intransigent, and then, second, how that understanding may help us to move in the direction of reconciliation in truth, for which Henson hoped.

And not Henson alone. It was the hope also of that bishop lying on his tummy in Jerusalem, writing that letter to his wife:

> There is a draught in the hut this evening for some reason, so my own candle is hopeless on the table. I have had to put it on the floor and write lying on the boards on my tummy.[5]

The bishop was William Temple, and he was writing, not only a letter to his wife, but also the draft of a major document. He was there in Jerusalem attending the conference of the International Missionary Council – the conference that was the successor of the great Edinburgh Conference of 1910, the beginning of the modern ecumenical movement.

The major document emerging from the Jerusalem Conference was an agreed statement on 'the Christian Message in relation to other religions'.[6] It was a ferociously difficult document to draft, simply because whenever

Christians of different kinds gather together, they are about as much in favour of agreement as two ferrets in a sack.

> 'The Apostle of Christianity,' Trollope once wrote, 'and the infidel can meet without the chance of a quarrel; but it is never safe to bring together two men who differ about a saint or a surplice.'[7]

That was quoted by Iremonger in his account of William Temple at Jerusalem where Temple was commissioned to do what he himself called 'his parlour trick' of finding words that all people could accept, even though individually they might have preferred something else. So he wrote in his diary on that same day, 3 April 1928:

> I was drafting all the morning, and seemed to be regarded as having done rather conspicuously my parlour trick of fitting everybody's pet point into a coherent document when they thought they were contradicting one another.[8]

So, was Temple the mere Maskelyne (or, to be anachronistic, the David Blaine) of ecclesiastical magic? Of course not – very seriously not. For him, as for Henson, the disunity of the Church was a blasphemy, a holding of God in contempt. In his *Readings in St. John's Gospel* he wrote:

> The unity of the Church is something much more than unity of ecclesiastical structure, though it cannot be complete without this. It is the love of God in Christ possessing the hearts of people so as to unite them in itself – as the Father and the Son are united in that love of Each for Each which is the Holy Spirit. The unity that the Lord prays that His Disciples may enjoy is that which is eternally characteristic of the Tri-une God.[9]

That is why Temple threw himself into endless conferences and committees in search of the fulfilment of that prayer – not just Jerusalem in 1928, but 'Faith and Order', 'Life and Work', Malvern and even Lambeth in 1930, where he worked passionately for the cause of what was called at that time 'home reunion'.

Even so, he recognized, as he put it, that 'the way to the union of Christendom does not lie through committee-rooms, though there is a task of formulation to be done there.' Where, then, does it lie? Temple had no doubt: 'It lies through personal union with the Lord so deep and real as to be comparable with His union with the Father.'[10]

It seems obvious. But in that case, why do Christians continue to hate each other so much – not maybe at the level of individual friendship, or of spiritual life, but certainly of institutional intransigence? And why are religious people so often so hostile to each other? Of course there are practical cooperations, there are Conferences and Parliaments of religions in which there is genuine affection and desire for a better way. And yet still religions (or at least some religious people) remain the agents of so many of the intransigent conflicts in the world. Along with that, they remain the agents of that kind of demonizing of their opponents which exacerbates religious conflict. It was grotesquely obvious in the history of antisemitism. It was seen also in the reaction of some to the announcement that Rowan Williams was to be archbishop of Canterbury. How could it possibly come about that those who have been brought into 'the glorious liberty of the children of God' (Romans 8.21) should make the word 'liberal' into a term of venomous abuse? Temple got it right: it is bad enough to be nasty to each other; it is, from what Christians regard as God's point of view, not even enough to be nice:

> Our friendships, our reconciliations, our unity of spirit in Church gatherings or in missionary conferences – beautiful as they are, and sometimes even wonderful in comparison with our habitual life of sectional rivalries and tensions, yet how poor and petty they appear in the light of the Lord's longing. Let all of us who are concerned in Peace Movements or Faith and Order Movements or 'Conversations' with fellow-Christians of other denominations, take note of the judgement under which we stand by virtue of the gulf separating the level of our highest attainment and noblest enterprise, from 'the prize of the call upwards which God gives in Christ Jesus' (Philippians iii, 14) – *that they may be one as we.*[11]

Given the paramount importance of unity and of reconciliation in Christian theory, why in practice do Christians fail to achieve them? And beyond that, why does the dialogue of religions turn out to be, beyond the obvious point of goodwill, a dialogue, not just of the deaf, but so often of the obdurately destructive? Given that our emotional and rational responses to the world around us and to each other are built in the same way, given that the phrase 'common humanity' now has a non-trivial content, why do Christians continue to hate each other so much?

Henson understood that question. So did Temple. And Temple glimpsed the answer to the question when he used, in his letter from Jerusalem, that word, 'coherent'. He produced, as he put it, a coherent document that people in disagreement with each other could nevertheless accept as their own. For it is in that word 'coherence' that the irreconcilable divisions of Christians among themselves are rooted – and not only the divisions of Christians, but the divisions of the religions of the world from each other as well.

Coherence

So what does coherence mean in this context of trying to understand why the divisions among Christians and between religions are so intransigent?

The basic point is simple: all religions operate on the basis of a coherence theory of truth, and they have no option but to do so. That may sound a bit esoteric, but the point is truly important. All religions put forward massive life-enlightening, life-saving claims about putative matters of fact: that God acts not just to teach but to be the way of salvation, that the Buddha teaches, not just to enact but to be the way of enlightenment.

But neither Christians nor Buddhists can produce God or the Buddha as objects among other objects in the universe. God is, to quote the first of the Thirty-Nine Articles, 'without body, parts, or passions';[12] the Buddha neither is nor is not, and it is a subversive distraction to ask where or even whether he is.[13]

Exactly the same will be true of many of the major claims advanced by religions: often what they are talking about cannot be produced as though the warrant for their proposals is a simple matter of inspection and verification.

How, then, are religions going to justify the claims that they make? Clearly not by holding up two hands, or holding up one God and one Buddha in front of their face. In other words, very few religious claims can be justified by that kind of empiricism – the kind described by Lonergan as 'going outside and taking a look'. It is that kind of empiricism which underlies *correspondence* theories, as opposed to coherence theories, of truth. To quote Ralph Walker's definition:

> The coherence theorist holds that for a proposition to be true is for it to cohere with a certain system of beliefs. It is not just that it is true if and only if it coheres with that system; it is that the coherence, and nothing else, is what its truth consists in. In particular, truth does not consist in the holding of some correspondence between the proposition and some reality which obtains independent of anything that may be believed about it.[14]

The contrast between correspondence and coherence theories is thus great. However, so what? No religion even seeks to justify its propositions or its actions by claiming some demonstrable correspondence between its sentences (or at least many of its important sentences) and what they purport to be about.

In contrast, where religions are concerned, what does justify propositions and actions is the extent to which they cohere with the system within which they are embedded – systems that include propositions, but include much else as well, such things as hymns, poems, art, rituals, prayers and certainly behaviour. In addition, any religious system that derives itself from revelation will, always, insist on checking out the extent to which propositions and actions are coherent with the scripture involved. Those who objected to the appointment of Rowan Williams as archbishop of Canterbury did so on the grounds that in their view he was not obedient to the Biblical message.

So far, so obvious: religions must of necessity rely on a coherence theory of truth. There is no choice about it. But the trouble is that there are immense problems with coherence theories of truth.

First and most obviously, if the requirement is simply that the system be internally coherent, then there will be many, perhaps infinitely many, different and incompatible systems of belief each of which is entirely coherent in itself. That is exactly the situation in which religions find themselves, and in which Christians also find themselves, since the many subsystems in which they live are passionately coherent, but they are incompatible with each other.

Thus relying on coherence *alone*, it would follow that the truth or otherwise of the propositions contained in a work like Hitler's *Mein Kampf* can be assessed only on the issue of whether the work is or is not internally coherent, and coherent also with the wider system of ideology in which it is embedded. Not even consequentialist considerations can intrude on a coherence theory of truth. Relying on coherence *alone*, all scriptures are equally true, 'the Vedas, or the Tiruvaymoli, or the Guru Granth Sahib, or Tanakh, or the New Testament, or the Quran, to mention only a few' (p 9).

Why must that be so? Because, since coherence is a matter of the internal relations between the components of the belief system, it does not depend in any way on the connection between the belief system and anything external to that system. It depends only on the coherence of the system. So it is coherently justified within a Biblical system to say that God created the world in six days, but how does that connect with propositions external to the system which suggest that evolution took a little longer? In terms of a Biblically literalist system of coherence, there is no reason why it *should* connect, and so we end up with creationism.

This has the curious consequence that a coherent system might be telling us a great deal about the world around us - for example, that it actually *was* created in six days - but how would we know? If no input is allowed from a world external to the system (e.g. from science), we would have no way of knowing whether what the system tells us is true or false in any other terms: if I say that the moon is made of green cheese, that will be entirely coherent in any system which maintains that all moons are made of green cheese. There is no way of challenging that unless we move outside the system altogether - exactly what the humanist and secularist critiques of religion have always done: they send (or rely on others to send) people to the moon.

What then makes the matter worse is that those criticisms cannot be convincingly answered from within a system relying on coherence, because its own arguments, being contained within the system, will be entirely circular: the truth of any proposition will depend on the truths of all the propositions in the system, which in turn depend on each other. It is that kind of circularity that occurs so often in religious dialogue.

Put as briefly and as crudely as that, all those objections might be answered, as indeed they *are* answered by, for example, Ralph Walker in his book, *The Coherence Theory of Truth*.[15] Even so, he still concludes that no pure coherence theory can succeed, because, as he puts it, 'no such theory can give an adequate account of what it is for a proposition to be believed, and hence to be a candidate for determining what the coherent system is.'[16]

However, the point here is not to pursue those arguments. It is, more simply, to observe that while religions have no option but to operate as coherence systems, coherence systems are fraught with massive problems. Unless those problems are recognized and addressed, there is no hope of delivering Christians from their disobedient disunity, or religions from their destructive mistrust of each other.

So what do we do? The very first necessity is to recognize that correspondence and coherence theories of truth are *not theories of truth*. Certainly they are spoken and written about in that way (as up to now in this book), but that is just a convenient (though inaccurate) shorthand. In fact, in the first instance, they are theories of epistemic justification.

How Do We Know What We Know? Epistemic Justification

Theories of epistemic justification are, quite simply, theories concerning how we justify what we claim to know (Greek *episteme*, 'knowledge'), the warrants we offer for the claims to knowledge that we make, the reasons why we hold the beliefs that we do. They are not, as such, theories of truth. The warrants we offer to justify what we say (I learnt it at my mother's knee; the Bible says) may supply the reasons why we have spoken or acted in a particular way, but they do not necessarily, in themselves, establish truth – even though a correspondence theory of justification is often believed to do so, because in its simplest form it holds that propositions are justified when they correspond to what they purport to be about: 'There are two hands in front of my face.' There is a correlation between words and reality by means of the signals that people receive from the world, and I am justified in saying that there are two hands in front of my face; and how can that not be true?

The answer is, because I might be mistaken. Perhaps I cannot count. Perhaps I am double-sighted. That is why, in contrast to a *correspondence* theory of epistemic justification, *coherence* theories claim that the foundations of knowledge in a real world external to ourselves are always open to the possibility of mistake and error. It is always the case that one might, by some further test or observation, be shown to be wrong. An appeal to the history of science shows how often that has happened. In the end, our propositions in such things as history or science, let alone theology, are vastly underdetermined by their correspondence to something external to ourselves. The most that can be allowed is some form of critical realism.

Critical realism means, roughly, that virtually everything we say, not least if we are scientists, is approximate, provisional, corrigible and often wrong, at least from the point of view of later generations. But (against deconstruction) it is at least wrong about something: there is sufficiently what there is in the case of the universe to act as a constraint on what we say, even though no one knows, incorrigibly and completely, 'what the universe is'. It is because there is consistent constraint and input from the external world (often in the form of conducive properties) that science achieves its reliability, while remaining at the same time corrigible.

Coherence theories of epistemic justification have no particular investment in critical realism, but they agree with its criticism of correspondence: in the end, our propositions are vastly underdetermined by their correspondence to something external to ourselves. So instead, coherence theories argue that the only warrant or justification that we can hope for in the case of any particular proposition is that it is coherent with the whole body of beliefs and propositions in the system to which it belongs.

To give an example: in January 2003, the fossil remains were reported of six four-winged, or four-flapped, gliding dinosaurs, *Microraptor gui*. The claim was then made that the wings of birds evolved from those gliding dinosaurs, instead of, as used to be thought, from later flying animals in trees with forelimbs adapted for flight. Clearly that new claim could not be justified by the direct observation of a series of events 110 million years ago. It can only be justified by the way it fits coherently into a whole network of propositions that make up the theory of evolution: it is coherent with them; and the whole theory of evolution is itself embedded coherently in a larger system of beliefs and propositions that make up the coherent story of biology. It is an important reminder that the natural sciences themselves rely a great deal on coherence for the justification of what they claim.

That equally is how religions also proceed. They cannot produce for observation God or the Buddha as Tathagata, nor can they produce for observation events in the past that are formative - indeed, transformative - of lives in the present. As a result, many of the propositions of religion are checked for acceptability by the extent to which they do or do not cohere with the system in which they are embedded. Are they sound, to use a common word? Do they conform to the magisterium of the Church, to use another common phrase?

That check for coherence is particularly obvious in religions that are derived from revelation, and it issues, in Christianity, in such natural consequences as systematic theologies, confessions, catechisms and creeds. The propriety of these can be checked by the extent to which they are coherent with the system as a whole, in which coherence with the (claimed) revelation is paramount.

However, what happens if some believe of others that their proposals are *not* coherent? They may, of course, burn them at the stake. Or, far more commonly, the stories drift apart into separate systems or really subsystems. The stories in these subsystems remain coherent in themselves, and they often claim to be the *only* story that is the correct, coherent version of the religion in question. In the case of Christianity, we get Methodist stories, Anglican stories, Catholic stories, and not just Catholic stories, but Opus Dei stories, Jesuit stories - or among Anglicans, Reform stories, Forward in Faith stories, and so on.

But so long as coherence is the *only* court of appeal, there cannot be any decision, or for that matter any reconciliation between them. They exist as closed circles of internal coherence. And that is what so many religions and so many subsystems within religions actually are, closed circles relying on internal coherence for the *justification* of what they say and do. They then make the disastrous and fatal error of assuming - or even of saying - that justification is the same thing as absolute truth.

But theories of justification are *not* theories of truth without further thought. It is only if something is allowed or is encouraged to break into the circle that a coherence theory of justification can move in the direction of truth. BonJour, for example, offered a defence of a coherence theory, in which he obviously rejected foundationalism (the view that incontrovertible foundations of truth can be identified), but at the same time he held fast to what he called 'a realist conception of independent reality'.[17] But that is only possible if the closed circle of coherence is in fact broken. In his own words, 'the apparent circle of justification is not in fact vicious *because it is not genuinely a circle*'.[18] He therefore argued that if a coherence theory is to give a fully explicit justification of a particular empirical belief, it will have to pass through four steps or stages of argument:

1. The particular belief must be inferable from other particular beliefs and further relations among particular empirical beliefs;
2. The overall system of empirical beliefs must be coherent;
3. The overall system of empirical beliefs must itself be justified;
4. The particular belief in question must then be justified by its coherent membership of the system.

It is still coherence that offers justification, but in Step 3 the theory accepts input from the world external to itself. Or to put it the other way around, if the theory wishes its proposals to move from fiction to fact, it must check into the world it purports to be about in ways that satisfy the criteria of empirical observation, even though its attitudes to those observations, and to reports made on the basis of those observations, remains one of a highly critical realism.

So the strength of coherence theories of epistemic justification is that they foster the telling and the elaboration of important stories – the story of evolution, for example, or even those grand narratives of which postmodernists are so suspicious – and of which, as we saw, Henson was equally suspicious. For he recognized the way in which the stories of Gore, or for that matter of the British Israelites (p 14), are coherent – they 'hang together' in themselves. But are they true? The weakness of coherence theories of justification is that they lack criteria within themselves for answering that question.

After all, on the basis of coherence theories one can write science fiction or science fact. What is the difference between them? In terms of a coherence theory of justification, none whatsoever. So what does make the difference?

Think again of the theory of evolution. It *has* to rely on coherence for the justification of much of what it claims, because no one can observe events 110 million years ago. But it does not rely on that alone. It moves in the direction of truth by allowing input from the world external to itself, the world of inorganic chemistry, for example, or of archaeology, so that it can move by way of analogy and of abductive inference from the observed world of the present to the unobservable world of the past.[19]

Science fiction is not a part of science in that way. It invents a world that cannot, by definition, *ever* have been experienced. Edmund Crispin made exactly that point in the very first sentence of his introduction to the renowned Faber and Faber series, *Best SF*:

> A science fiction story is one which presupposes a technology, or an effect of technology, or a disturbance in the natural order, such as humanity, up to the time of writing, has not in actual fact experienced.[20]

Nevertheless, any science fiction story that hopes to evoke a judgement of value, a judgement that the story is good, will aim to be internally coherent: it must have the internal coherence that is sought, for example, in films under the name of continuity – sought but so often missed.[21]

But internal coherence is not enough for a science fiction story to evoke a judgement of approval. It will aim also to be coherent with the world of actual science at the time of its writing. Because the closed circle of the story is broken, and because the internal coherence of the story is allowing, not just input but also constraint, from something external to itself (the external world of the natural sciences), science fiction is not just a collection of good stories, but can be a serious exploration of what can truly be said about ourselves.

The same is true of myth. Myth might be just a good story relying on coherence alone. But the power of myth is that it moves in the direction of truth by connecting itself to worlds external to itself. Tolkien knew this very well. When he wrote *The Lord of the Rings*, he took immense care to create a coherent system – so much so that many of his letters are taken up with answering challenges to the coherence of his creation.

One letter, from Miss Beare in 1958, was set with numbered questions like an exam paper: Question 4. Explain the meaning of the name Legolas. Question 1 probes the internal coherence of the narrative: 'Why (in the 1st edition, I.221) is Glorfindel's horse described as having "a bridle and bit" when elves ride without bit, bridle or saddle?'[22]

Tolkien answered those questions of coherence with very detailed care. But internal coherence was not enough. He also claimed for his enterprise the word 'myth', by which he meant that, although not all the details of the story would 'check in' to any real world of the past, on the other hand his story does 'check in' to the real world at some points. So he wrote to Miss Beare:

> May I say that all this is 'mythical', and not any kind of new religion or vision. As far as I know it is merely an imaginative invention, to express, in the only way I can, some of my (dim) apprehensions of the world.

That is why Tolkien was able to explore themes that are important, not just to himself, but to us all. In his own words, he constructed an imaginary time, but kept his feet on his own mother-earth for place:

> I have, I suppose, constructed an imaginary *time*, but kept my feet on my own mother-earth for *place*. I prefer that to the contemporary mode of seeking remote globes in 'space' [by which he meant C.S. Lewis]. However curious, they are alien, and not lovable with the love of blood-kin. ... Theologically (if the term is not too grandiose) I imagine the picture to be less dissonant from what some (including myself) believe to be the truth. But since I have deliberately written a tale, which is built on or out of certain 'religious' ideas, but is *not* an allegory of them (or anything else), and does not mention them overtly, still less preach them, I will not now depart from that mode, and venture on theological disquisition for which I am not fitted. But I might say that if the tale is 'about' anything (other than itself [which would have kept it a closed circle of internal coherence]), it is not as seems widely supposed about 'power'. Power-seeking is only the motive-power that sets events going, and is relatively unimportant, I think. It is mainly concerned with Death, and Immortality; and the 'escapes': serial longevity, and hoarding memory.[23]

From all this it follows that the strength of coherence theories of epistemic justification is that they allow the telling of many stories. Their weakness is that they offer *on their own* no criteria outside themselves for distinguishing between them in terms of fact or fiction. On the basis of coherence theories of justification, how can we tell whether Christian claims are true? How can we tell the difference between science fiction and science fact? And how do we evaluate the differences between Christians of different kinds, and the differences also between religions?

Closed Circles and Coherence

The answer is of course that we *cannot ever do so* if they rely on coherence alone to justify what they say, because they will remain closed circles. They can only move in the direction of truth if they accept that their relation to the worlds external to themselves is a necessary part of what makes them not only coherent but also true.

If, conversely, a religious or a Christian system remains a closed circle and refuses to allow any challenge or confirmation from the worlds in which it lives, then we have no way of telling whether it is fiction or fact. It will remain profoundly coherent, and for that reason it will often be fierce, even violent, in defence of the boundary of its circle. There is no doubt that people can live confidently and happily in the closed circles of coherence that religions are, and that their lives will often be characterized within those circles in impressive ways. But they cannot claim to be true beyond the boundary of that circle.

It is easy for those who make claims for the truth of any particular religion, for example in Christian evangelism, to forget this: they rely on a coherence theory of justification, but then presume that that is the same as a theory of truth. But it is not. If truth is to enter into the matter, all the stories that humans tell, whether Christian or Buddhist or scientific or historical, or of any other kind, economic, geographical, and so on, must indicate how they check into what they purport to be about. All those stories may still rely on a coherence theory of epistemic justification, as clearly, to some extent, they all do, and it is the error of foundationalism to suppose that they do not.

But coherent systems cannot therefore dispense with the input from the external world as a constraint over what they say: it was the error of *unremitting* idealism to suppose that they can. Somewhere between those two extremes of foundationalism and idealism, there must be a better road to travel. Or so at least Henson believed.

That is why the topic of the lectures named after him remains so important: the appeal to history as an integral part of Christian apologetics, not, as Henson hoped, in the sense of being an appeal to incontrovertible facts, but as the constant willingness to 'check in' to the worlds it claims to be talking about. Since it is not believed that God is an object like a universe, still less an object in the universe, to be produced for some kind of observation, it follows that claims involving God will be relying on a coherence theory of justification. But if they have anything to do with truth, if, that is, they move beyond being systems that rely on coherence alone for the justification of what they say, then they have to indicate how they connect with constraint derived from outside the system, or how what they say about the observable universe (including within it the nature and purpose of human life) 'checks in' to what their statements purport to be about.

So, for example, the Christian story is tied into the world external to itself, not only of the present, but equally of the past, equally also of the future. If Christianity has anything to do with truth, if, that is, it moves beyond being a system that relies on coherence alone for the justification of what it says (often expressed in the form, 'the Bible says'), then it has to indicate how it checks into what it purports to be about.

Thus the claim that Jesus is the Son of God, who died on the cross to save us from the deserved penalty of our sins, is easily justified by the way in which it is embedded coherently in the whole system of Christian belief. But is it true? How does it check into the world external to itself –

the world, for example, in which Jesus lived, or, to give another example, the world and the way in which his followers live now, or to give another example, the many other ways in which human nature is understood? It is important here not to revert to the fallacy of the falsely dichotomous question (pp 77–8), *either* the internal coherence *or* the external checks, because *both* are needed and neither will do the whole work of establishing warrants for our assertions on its own.

It is the importance of *both* that Henson knew to be important. His own belief in the Incarnation was profound, as he said in the letter already quoted (p 16). But for that claim to be true, it must never evade the external challenge of historical truth. Of course others, like Gore, might argue against that, and they might raise petitions to get Henson deprived of his orders. They might, for example, argue on grounds of historical evidence. But in doing *that*, they will have accepted, epistemologically, that the circle is *not* closed: it is allowing input from the external world, in this case of historical evidence and argument.

If, on the other hand, they argue solely on grounds of coherence, as the Vatican authorities are inclined to do when they investigate someone who is accused of denying the teaching authority of the Church, their verdict can only be persuasive to those who are already inside the closed circle, and who accept that coherence is a sufficient condition of truth *on its own* – a major reason why the Gospel loses credibility in the real world, i.e. in the world external to the system. The closed circle will be coherent, its sermons powerful, its adherents loyal. But if it refuses any input from the world external to itself then no apologetic is possible, because there is no way in which we could know whether its claims are true or false. Unless evangelism or Christian apologetic understands the difference between coherence theories of justification and theories of truth, and why *both* are needed, it will be confined, as it so often is, to negotiating, or even engineering, a fit between the story and the hearer.

Henson saw that. He accepted with, as Gore recognized, 'good faith', the profoundly coherent Christian story, the story that moves people to conversion in all that it offers, moves them to mercy, moves them to anger, moves them to prayer in what it mediates of God. There *are* conducive properties in that story which engage people in their emotions, feelings and rationality, so that whereas before they were blind, now they see.

But however profoundly coherent that story is, and however much it must rely on coherence in order to justify what it says, it cannot rely on a coherence theory of justification alone if it wishes to make claims to truth.

Christian apologetic must build bridges between itself and the world in which it lives, the world of the present *and* of the past. It must show the effect of that story on lives and institutions through time and in the present, for worse as also for better: by their fruits you will know them.

That input – from history, from aesthetics, from ethics, from science, from the very lives that people live – is the only way in which that decisive shift can be made, from coherence as a theory of epistemic justification to coherence as witness to God who beyond all stories is the source and sustenance of all that is.

Without it, Christian claims remain powerfully justified by the way they hang together as a coherent story, but that is all. There can even be systematic theologies provided they recognize that if they rely, as most of them do, on a coherence theory of justification *alone*, they are of interest within the system alone – a reason why, as it used somewhat mockingly to be observed, the works of Karl Barth are rarely found on the shelves of probation officers or social welfare workers.

The point is that Barth's work is of interest only within the closed circle, where it is tested for approval or disapproval (i.e. for justification) by the extent to which it is coherent within the system. Barth himself would surely have resisted any other check on its truth, because that would have been to subordinate God to the measure of human enquiry: it would have been to subordinate revelation to reason.

And yet, in terms of Gospel and of evangelism, in terms of Christian apologetic, in terms of claiming the world and all its systems for God who has bestowed everything on us as gift, more is needed. Christianity, to paraphrase Tolkien (p 130), must keep its feet on God's own mother-earth. For that reason it has no option but to do the hard and laborious work of making connections, of 'checking in' to all those many worlds in which we live – the aesthetic world, for example, the moral world, the world of the sciences, the world of history, the world of suffering.

An Emotional and Rational Faith

In this book, an attempt has been made to show how Christianity does connect with some of those worlds in ways that Henson, in his day, could not possibly have imagined. Far from subordinating revelation to reason, it is exactly the other way around. Revelation comes to reason demanding that it takes seriously, as a matter of fact, the true and magnificent opportunity that exists within this human nature to be at one with God. There *are* conducive properties that lie within Scripture and within the worlds that humans inhabit which evoke and sustain that union with God.

But does that make 'God' simply an emotion? Of course not. God is mediated to us through the occasions that conduce the response of recognition and faith, but God is not identical with those occasions, since otherwise we would be back to saying that there is after all a single property of beauty or of goodness or of God that all must see like the redness of a book.

We do not *see* God as an object, still less as a property of some sort. Conducive properties reach us in ways that evoke and sustain the recognition of God as one who takes the initiative to reach us, in what all religions understand as 'grace', however different their words for that initiative may be. We do not create the conducive properties. They create us. Or rather, God, through them, creates us as a new creature, made much more than we could otherwise be. And in response, there are occasions when, to adapt the quotation from Bill Bryson (p 49),

> our minds, unable to deal with anything on this scale, just shut down, and for many long moments we are without speech or breath, but just a deep, inexpressible awe that anything could be so vast, so beautiful, so silent.

There are occasions when our rationality and emotion and our whole body can only fall down and murmur, 'My Lord and my God'. But even that is not compelled or coerced: it is a response of faith, because the entailment is not so strict that it leads coercively to only that one outcome. It did not do so for Bill Bryson – or at least he did not include it as part of his memory of that experience.

But by the route of epistemic entailment, the conducive properties of these occasions do lead to the abductive inference of God as the necessary source of them. They lead also to what Auden called 'a *super*-natural sympathy'. From the ordinary moments of our humanity, we are conduced, led, not to see God as an object among other objects, but to the recognition that God must be. For Auden, the occasion of which he was writing when he created that phrase, 'supernatural sympathy', was erotic and carnal love, in which we surely can, as he put it, 'find our mortal world enough'. True. But in those and many other occasions of our ordinary humanity, our emotions and rationality can be led to see and be seized by those transcendent universals that point unremittingly to God who is and who endures when 'beauty, midnight, vision dies'. Then, to quote the poem:

> Soul and body have no bounds:
> To lovers as they lie upon
> Her tolerant enchanted slope
> In their ordinary swoon,

Grave the vision Venus sends
Of supernatural sympathy,
Universal love and hope.[24]

The conducive properties through which God reaches into our humanity may be – often are – extremely simple. But they are real in effect, and they have consequence in changing the emotional and rational way in which we live. They lead to the life-characterizing metaphors through which God remains independent and yet invites, initiates and creates our union with himself.

Metaphor does not claim to describe God, still less to say 'what God is like': metaphors, in this case, are not similes. They explore what is believed to be true but which cannot be described in direct or resemblance terms. God is not *like* any thing: Deus semper maior: God is always greater.

But the life-characterizing metaphors of religion work within us, and they carry the biography of our own exploration and faith far beyond its point of departure. That is why religious metaphors hum and buzz with such vitality that they not only characterize our lives, they pour out into art, and dance, and drama, and prose, and poem, and praise. They enable us, if not to tread on holy ground, at least to know when to take off our shoes.

It is a new and far more splendid anthropology than the *divide et impera*, divide and rule, anthropology of the last 2,000 years, separating reason from emotion. Furthermore, it connects profoundly with the paradigmatic anthropology claimed in the New Testament, in which the human nature of Jesus was conjoined with God.

With *God*: a reminder that there remains the One who is, by definition, independent of the coherent stories that religions tell, while being their source and sustenance. The independence of God has to be defined in that way, since otherwise all that is associated with the concept of God cannot function. For God to be, God must be whether there are Jews, Hindus, Christians, or anybody else to say so. That in itself is not a conclusion about whether or not the word 'God' has reference to that which is the case beyond the definition. Thus the question of the reality of God, independent of this or any other world, and therefore certainly independent of the religions that nevertheless respond to God and embrace God in the stories they tell, remains to that extent open.

However, for the many theistic religions, the question of God's existence does *not* remain open. So how do their very different theistic stories 'check in' to what they purport to be about, the independent reality of

God? In those religions, that connection with God is understood as being a matter of response, so that there are in every religion fundamental ways in which people can, so to speak, 'check in' to what their beliefs purport to be about, in such ways as meditation, contemplation, prayer, worship, and also in action, not least in the love of their neighbour. Many of those 'ways' are described in my *God: A Brief History*. In their practice and unfolding development through time, they reinforce and confirm the coherence of each story in question. But how also do they relate to each other?

The Relation between Religions

That takes us back at once to the opening questions: how do the different and superbly coherent stories that constitute the different religions relate to each other? And how might the different systems within the boundary of Christian coherence engage with each other? Or to put the question more personally, does it make any difference into which one of these many stories a person enters?

On a coherence theory of justification, it cannot do so, because that is the consequence of relying on epistemic justification *alone*; and unless these stories are prepared to connect with the external world in the ways that I have tried to indicate, they must logically coexist as multiple worlds of equal validity. In each closed circle, a *choice* has been made about the reasons that are allowed to have validity (exactly as Miller described in a more general way: see p 6), and the choices have a restricting and limiting effect. Of course, each circle will give an account of how it is related to others, and that may well include an account of why it is the only way to salvation or enlightenment. But in themselves, those accounts can only be evaluated (as they are) by their coherence within the closed circle. They cannot be evaluated in terms of truth beyond the justification they seek within the closed circle.

If, however, these stories do allow input from each other or from the worlds they purport to be about, then very real distinctions do open up. There *is* a difference between Gore's ecclesiology and the British Israelites, let alone between Hindus and Muslims, Buddhists and Jews, or any other combination of religions.

Do the differences matter? Clearly they do in Gujarat or in Ramallah or in Kaduna or in the Sudan or in any of those many other places where the differences explode in violence. So how, if at all, do we learn to live together? And how do those who seek to be an answer to the prayer of Jesus, that they may be one, do something about it?

The precondition of any quest for union or reconciliation among Christians or for a less hostile relationship between religions must begin with a shared and explicit understanding that coherence theories of justification are not the same as theories of truth – and with a clear recognition also that coherence theories of justification cannot be converted into theories of truth simply by the declaration that that is what they are – because those declarations are made within the systems, and are entirely circular. James Mumford, in his seventeenth-century *The Question of Questions: ... Who Ought to be Our Judge in All These Differences?*,[25] had no difficulty in justifying his answer, the infallible Church, and in concluding (the title of his last chapter), 'The Roman Church having been proved to be our infallible judge, all under pain of damnation are bound to submit to her judgement.' But because the justification is established entirely by its internal coherence within the system, it has not yet, in the intervening centuries, convinced all the Protestants in Belfast.

The next step, therefore, must be to explore the extent to which any input from worlds external to those systems is or is not allowed, since if it is not allowed, then questions of truth cannot be addressed. Questions of the relationship between closed systems certainly can, but they will be questions of whether coexistence or cooperation is allowed on the terms obtaining in the respective systems, and if so of what kind.

Thus much has been made in recent years of the new alignment of Christendom, no longer the old divisions between Protestants and Catholics, but a new alliance of conservatives among both Protestants *and* Catholics over against liberals of any kind. But in fact where religions or subsystems within religions are concerned, the real division is between open and closed systems of coherence, since there can be conservatives and liberals in both. Examples of closed and open subsystems within a larger religious system are Orthodox and Reform Jews, Hanbalite and Hanafite Muslims, adherents of Swaminarayan and of Sri Aurobindo, Nichiren Shoshu and Friends of the Western Buddhist Order.

Yet even from those few examples alone, it is obvious that no system is completely open (since it would otherwise disintegrate at once) or closed. Islam is an example of a closed system because it believes itself to be derived from the final, complete and universal revelation from God, illustrated in an authoritative way in traditions (Hadith) of the words and actions (and silences) of the Prophet and his Companions. Islam is not open to challenge or correction from the world outside itself. In an interview recorded in *What Muslims Believe*, I asked a Muslim whether Darwinian theory would have to be regarded as false if it conflicted with the Quran. He replied:

Certainly those parts that conflicted with the Qur'an would be false if they involved something basic and fundamental, and not just an interpretation (we must be clear that what we are saying is 'from the Qur'an') - but if there were something in the Qur'an which contradicted what Darwin has said, then we would not accept the theory, but we would accept the fact, and that is the Qur'an.[26]

Yet clearly Quran and the Hadith do not deal with all circumstances that people encounter, so that schools or subsystems were developed of Sharia, in which effort, *ijtihad*, was made to interpret and apply the tradition or Sunna. Various methods came into being, of which the most important were *ijma'* (consensus), *qiyas* (analogy), *ra'y* (considered and informed personal opinion) and *istihsan* (contextual opinion, where strict analogy is set aside in the interests of justice or public benefit).[27] Those who adopt all four of those methods open the system to the world far more than those who do not. The Hanbalites go furthest in the direction of closure by allowing only *ijma'*, and even then restricting it to the generation of the Companions of the Prophet and their generation. Only the Wahhabis can go further than furthest, by seeking to live as in the earliest days of Islam.

In general, therefore, the difference between 'closed' and 'open' is a matter of emphasis, but even more of intention. However, where the intention of leaders and participants in any system is to keep it closed, then only questions of coherence, not of truth, can be explored. Thus, to give an example, the Christian World Congress on Evangelism was held in Berlin in 1966. One of its sections was devoted to the issue of universalism (the issue of whether all people will be saved, or whether only those will be saved who have made a confession of faith and commitment to Christ before they die). Four speakers argued against universalism on the grounds that there is no sufficient Biblical warrant for it. Thus they had no option but to argue against it, because they were operating in a closed system relying on internal coherence for the justification of what can and cannot be said with approval. After the four papers, there then followed a group discussion which was reported as follows:

> The discussion was opened by a delegate from Israel, a man concerned with the work of Hebrew Christian missions. He spoke of his father and mother who were strict Orthodox Jews. He said that they fasted and tithed and kept scrupulously all of the Commandments, that no evangelical witness was ever given to them, that the only Christian Church they ever saw was one in which idols were worshipped, a matter which would repel any Jew. He said his parents were put to death in the slaughter of the Jews by Hitler. He had himself come to faith in Jesus Christ as the Messiah and his Savior. He asked the question very movingly, 'Is there no hope for my parents who died, as far as I

know, without any faith in Jesus Christ?' The members of the panel said that such questions had to be dealt with not only from a theological standpoint but from a pastoral standpoint, that the Judge of all the world would do right, and that one must understand the hardship through which these people had passed, that some questions could not be completely answered in this life.

There were those present who immediately responded that we had no authority to go beyond what Scripture states, that there is no apparent hope for anyone except through faith in Jesus Christ.[28]

In terms of a closed system looking for justification from within the system (in this case, from the Bible literally and uncritically understood), it is obvious that the second response 'from those present' was justified and was correct within the system. The hesitations of the panel show how reluctant they were to pursue coherence with such consistency that it ends up with a characterization of God as some kind of monster. As I put it in *God: A Brief History*:

> It is easy to construct from the Bible a picture of God as a Mafia boss – a literal God-father: he has his own family and protects it fiercely, especially where marriage is concerned; he has his own territory and protects that also; he expects certain, very specific, standards of behaviour (eventually, in this case, written down, but for a long time an unwritten code); he engages, if necessary, in war with rivals; and deals with offenders in a brutal and conclusive way. Like Mario Puzo's Godfather, God makes offers to the Israelites that they can't refuse.[29]

Of course, that is not the only characterization of God in the Bible. Indeed, in the very next sentence I wrote, 'But that is not the only or the whole picture.' That raises at once the issue of history and hermeneutics (of how the Bible is to be interpreted). The books of the Bible were written over a period of more than 1,000 years, in which the characterization of God changes greatly, as the people involved came to understand more profoundly the One with whom they had to deal. This means that the Bible itself contains a record of change and correction. It therefore betrays the nature of the Bible understood as revelation to treat it as though it is a builder's yard from which one can take out what O'Donovan called 'moral bricks', as though all bricks are the same – as though texts can be extracted from any part of Scripture as discrete items, without reference to context. To do that is to ignore the fundamental difference in the Bible between commands that are domain-general and context-independent, and applications that are context-dependent even when they appear in the form of rules (pp 82–4).[30] What Scripture will reveal to us is the kind of wall worth building which will evoke the predication of 'good'. As O'Donovan put it:

> The items in a code stand to the moral law as bricks to a building. Wisdom must involve some comprehension of how the bricks are meant to be put to-gether. This has an immediate bearing on how we read the Bible. Not only is it insufficient to quote and requote the great commands of the Decalogue and the Sermon on the Mount (and there are still many who need persuading of this in practice if not in theory); but it would be insufficient even if we added to them, if we could compile a complete list of things commanded or prohib-ited; it would be insufficient even if we included in such a list, with a shrewd awareness of the relativity of semantic terms, principles derived from other modes of moral teaching in the Bible, such as stories, parables or laments. We will read the Bible seriously only when we use it to guide our thought towards a *comprehensive* moral viewpoint, and not merely to articulate disconnected moral claims. We must look within it not only for moral bricks, but for indi-cations of the order in which the bricks belong together.[31]

As O'Donovan immediately realized, there may be, as he put it, 'some resistance to this ... from those who suspect that it will lead to evasions of the plain sense of the Bible's teaching'. Within a closed system that uses the Bible as a God-given and God-protected Book, all parts will have equal authority when it comes to justifying an assertion. From the Bible used as a warrant in that way it has been possible to justify the burning of witches, the shipping of Africans into slavery, the treatment in law of women as 'a kind of infant', the use of animals to serve human ends, and the fighting of wars. Jews and Christians read passages from Scripture that celebrate the way in which God ordered and helped his chosen people to destroy the inhabitants of the Promised Land:

> He [the Lord] it was who struck down the firstborn of Egypt,
> both human beings and animals;
> he sent signs and wonders into your midst, O Egypt,
> against Pharaoh and all his servants.
> He struck down many nations
> and killed mighty kings –
> Sihon, king of the Amorites,
> and Og, king of Bashan,
> and all the kingdoms of Canaan –
> and gave their land as a heritage,
> a heritage to his people Israel. (Psalm 135.8–12)

Constant repetition of these 'words of God' internalizes them as a power-ful constraint over behaviours and attitudes in the present. The political consequences of this in the Middle East are tragically obvious. In a brilliant analysis of the far-reaching consequence of this, Jill Hamilton has shown how the politicians who brought into realization the Zionist dream of a Jewish homeland in Palestine were mainly Bible-based Protestant Christians.[32]

It follows that the recovery of the Promised Land and the destruction of those who stand in its way are entirely coherent in a closed system relying on an uncritical use of the Bible to justify its claims. There are both Jews and Christians who exemplify that position. For sure, there are other Jews and Christians who do not live in that closed system, and who realize that since history was not overruled or ignored by God in the way that Scripture came into being, it cannot be ignored by them when they appropriate these words into their lives as a constraint over what they do and say (especially if they are Christians, since otherwise they would trivialize the Incarnation).

So the issue of history and hermeneutics is not in the least remote or trivial. The issue is the extent to which a religion (or subsystem within a religion), relying as it must on coherence for the justification of its claims, can *only* function as a closed system. Religions are, in this respect, very different. Muslims, for example, do not face the same issue of history in relation to the Quran as Christians do in relation to the Bible, because the claims that the Quran makes about itself and its relation to history (claims that can only be justified from within the closed circle of that system) are different. It is not surprising that the political consequences when such systems are in conflict are so intractable.

There is no question that closed systems (especially those that are long running) generate immense loyalty and confidence among those who belong to them, because they know the terms and conditions of the coherence that justifies their allegiance. By insisting on the terms of inclusion, closed systems usually end up emphasizing exclusion. This consequence can be seen in the opening words of the Constitution of the First Baptist Church of Berwyn, Pennsylvania:

> We disavow the position of the World Council of Churches, the National Council of Churches, the National Association of Evangelicals, and any other association or fellowship that would be in sympathy with them. We stand in opposition to the Ecumenical Movement, New Evangelicalism, Inter-denominationalism, Protestantism, Neo-Orthodoxy, and cooperative Evangelistic programs between churches and people not alike of precious faith.[33]

On the basis of coherence, it is easy to determine which religious system or subsystem is true to the exclusion of others. The question 'Is one religion as good as another?' is the title of a book by John MacLaughlin.

Since by 'religions' he meant, not the religions of the world, but the subsystems of Christianity, his question was easy to answer on the basis of coherence: first, the marks of true religion are derived from the system: the true religion or Church, he claimed, must exhibit unity, universality and infallibility; then, second, the reader is invited to decide which religion or Church exhibits those characteristics:

> Well then, dear reader, raise your eyes, look around you, inquire, examine. Is there any Church on earth in which this unerring authority is found – in which this infallible voice speaks? Where is it? Which is it?[34]

The expected answer is, of course, 'In Rome'. But another answer could have been, 'In Japan'. MacLaughlin's book was published in 1887, and reprinted in 1891. In 1889 the Meiji constitution was adopted in Japan, and in 1890 the Imperial Rescript on Education was distributed to all schools in Japan and hung up beside the portrait of the Emperor, to both of which obeisance was made. The Rescript states:

> The Way here set forth is indeed the teaching bequeathed by Our Imperial Ancestors, to be observed alike by Their Descendants and the subjects, infallible for all ages and true in all places.[35]

Here the infallible voice speaks not in Rome but in Tokyo.

It is obvious therefore that any system which relies on coherence for justification, and which identifies coherence with truth, can put itself at the centre of its own circle. It can then evaluate others who are by definition 'outsiders'. The evaluation does not have to be as negative as that of MacLaughlin. The encyclical *Ecclesiam Suam*, issued by Pope John XXIII just before the Second Vatican Council, was intended to be more positive. It adopted this methodology 'by describing a series of concentric circles around the central point at which God has placed us' (§107). The outermost circle comprises the entire human race, the next comprises Jews, Muslims and the followers of the great Afro-Asiatic religions, and the next other Christians, 'the circle which is nearest to us' (§109).

It is a clear statement of the relation between religions: put the Pope at the centre ('the circle which is nearest to *us*'), and then assess other religions on the basis of the extent to which they accept the centrality of the Pope. The same methodology is open to all closed systems, but they of course would put something or someone different at the centre. They all produce exclusivist accounts of the relation between religions, or within religions.

However, the Second Vatican Council, while adopting the same methodology, made a decisively important correction to *Ecclesiam Suam*: it put

God in the centre of the circles of religion, not the Pope. That produces an inclusivist account, because it accepts that other religions may be centred on God, even though Roman Catholicism remains the true account of the matter (truth being determined by coherence within the Catholic system). At best, therefore, those outside may now be regarded as 'anonymous Christians'. Beyond that there may be pluralist accounts, in which religions are regarded as equally valid in relating their adherents to that which is ultimately real – or Real – to which religions give different names. This was John Hick's Copernican revolution. Just as Copernicus removed the earth from the centre of the universe and replaced it with the sun, so this Copernican revolution removes the claim of each religion to be the centre, and replaces it with the ultimately Real, however that is described in approximate and corrigible human languages. So he argued:

> The great post-axial traditions ... exhibit in their different ways a soteriological structure which identifies the misery, unreality, triviality and perversity of ordinary human life, affirms an ultimate unity of reality and value in which or in relation to which a limitlessly better quality of existence is possible, and shows the way to realise that radically better possibility. ... Thus the generic concept of salvation/liberation, which takes a different specific form in each of the great traditions, is that of the transformation of human existence from self-centredness to Reality-centredness.[36]

The problem with this particular 'Copernican revolution' is that even if the model is accepted, of 'planets circling around the sun' from which they all receive their heat and light, it does not follow that all planets are equally able to sustain life. If it is possible to be religiously right, it must be possible to be religiously wrong.

Exclusivism, inclusivism and pluralism are all options that those within closed circles can choose.[37] Because of the problems in each of them, I have argued for a fourth option which, for want of a better word (and there must surely be a better word), I have called 'differentialism'.[38] Any of these options can be chosen by those who live in closed systems, and the exploration of how religions should live together will be an exploration of these options.

However, if inputs of constraint are allowed or even welcomed from worlds external to the system, then it becomes imperative in this exploration to value diversity (as inclusivism, pluralism and differentialism can in fact do). Unity is not uniformity. If we know anything of God from the analogy from earth to heaven, then we know that God, in creation, is one who loves diversity: every snowflake is different, the iris of every eye is different, the earth (small though it is) contains about 30 million different plants, animals, fungi and protists,[39] the near infinite ways in which the genetic code can be written creates the brilliant diversity of organic life.

On the basis of diversity, it can then be seen why circles *to some extent* must be closed. It is easy to see why closed circles can be dangerous, as indeed they are when they confuse internal justification with truth. That way lies the closed and fanatical mind, what MacMillan called 'the sealed enclosures of sectarian certitude'. They are frightening enough in the religious world. They are even more alarming in the political. The man who nearly persuaded Eisenhower that atomic weapons should be re-garded as conventional weapons of war, and whom Churchill described, after meeting him in 1953, as a terrible handicap and as preaching like a Methodist minister whose 'bloody text is always the same, that nothing but evil can come out of meeting with Malenkov',[40] was accused by Rein-hold Niebuhr of 'fanatic oversimplification', the equivalent of MacMillan's 'sealed enclosures of sectarian certitude':

> Mr. Dulles' moral universe makes everything quite clear, too clear. Yet it does not illuminate any of the problems created by the Russian economic advances both at home and in Asia and Africa. And it does complicate our relations with our allies who find our self-righteousness very vexatious. For self-righteousness is the inevitable fruit of simple moral judgements, placed in the service of moral complacency.[41]

But the paradox is this: we have to remember what biology has taught us, that constraint is the necessary condition of freedom, and that to some extent the process of energy and information requires boundaries. Con-straint sounds a restrictive word, as indeed it may be: we try to constrain those who are a danger to themselves or to others. But constraint is also the necessary condition of freedom. That was the argument of Stravinsky (to take only one example) in his Harvard lectures on 'the poetics of music': without boundaries and constraints, a composer is let loose into a paralysing infinity of equal possibilities:

> At the very beginning of my course, I gave notice that I would continually come back to the necessity for order and discipline. As for myself, I experience a sort of terror when, at the moment of setting to work and finding myself be-fore the infinitude of possibilities that present themselves, I have the feeling that everything is permissible to me; if everything is permissible to me, the best and the worst, if nothing offers me any resistance, then any effort is in-conceivable, and I cannot use anything as a basis, and consequently every undertaking becomes futile. ... So here we are, whether we like it or not, in the realm of necessity. And yet which of us has ever heard talk of art as other than a realm of freedom? This sort of heresy is uniformly widespread because it is imagined that art is outside the bounds of ordinary activity. Well, in art as

in everything else, one can only build upon a resisting foundation. ... My freedom thus consists in my moving about within the narrow frame that I have assigned myself for each one of my undertakings.

I shall go even further: my freedom will be so much the greater and more meaningful the more narrowly I limit my field of action and the more I surround myself with obstacles. Whatever diminishes constraint, diminishes strength. The more constraints one imposes, the more one frees one's self of the chains that shackle the spirit.[42]

The paradox extends to all human creativity, and certainly to religious and spiritual life: the more elaborate the constraints are, the greater the resulting freedom is for possible behaviours. It was this that created Ashby's principle in cybernetics: 'Where a constraint exists, advantage can usually be taken of it.'[43] That is exactly what we observe in the case of biology: the more constraints there are, the greater the degrees of freedom for the organism in question. This is the liberating consequence of constraint. Humans are constrained by much more than is a rockweed kelp or an amoeba, and as a result they are set free to do and to be much more than a rockweed kelp. It is only humans who in consequence can heed the cry of the poet Andrew Young:

Up, valiant soul, put on thy jumping shoes
Of love and understanding. Soon I should learn;
The ghost must go; the cocoon, spun by the worm,
The butterfly would burst. New eyes would see
The invisible world into which my brother vanished.[44]

This is the puzzle that religions have in general failed to solve so far. The circles of religion are circles of constraint, in which people learn how to live, how to pray, how to meditate, and so on, and the boundaries of the system are therefore necessary. However, maintaining the system and its boundaries becomes so important for some people (who often call themselves 'religious leaders') that it becomes an end in itself. That is a major reason why religions are capable of producing lives of sour intolerance and bigotry: in any barrel of apples, it is certain that there will be a few rotten ones; it is, however, an entirely different matter when the whole barrel seems to be designed to produce rotten apples, and that is by no means unknown in religious history, past and present.

Yet the point of the boundaries and of constraint is to be not an end (maintaining the system) but a means to a far greater end, the composing of the symphony, the singing of the song, the attainment of God. When people internalize the relevant constraints, they are set free to break out of the circle and find God – God who is greater (Deus semper maior) than the circles of constraint which (paradoxically) enable people, sisters *and* brothers, to see the invisible world. My book *God: A Brief History* is a record and celebration of many of them, from all parts of Christianity, and from all religions, jumping out of constraint into the condition of love, which is, as Temple said, the interior nature of God.

This means that the ecumenical quest (to which Temple committed himself, pp 121–2) cannot succeed if one closed circle seeks to absorb all others into itself on its own terms. But it cannot succeed, either, if it seeks to abolish the circles that have opened to those within them the vision and the attainment of God and of the many other goals in lives that evoke the predicate of 'good'. The ecumenical quest so far has, in general, held out unity as the goal. In fact it should hold out diversity as the goal, in order to explore how the circles can be, without losing their constitutive attributes, no longer entirely closed but open to truth from worlds external to themselves, open to each other without losing their own identity, open above all else to God.

So the coherent circles are important, and they must be valued, despite the fact that they can be, and often are, immensely destructive when people become defensive of their circumference and 'take on' the world outside, in defence or in aggression. C.S. Lewis used to speak of 'the sin of the inner circle'.[45] Equally threatening is the sin of the outer ring road, converting a means of communication into a defensive and impermeable boundary. But when constraints are understood positively, as the condition of freedom and also as the wiser way to liberalize our understanding of 'cause' (pp 4–7), then the concepts of constraint are immensely helpful in trying to move the relationships between religions and within religions from threat and hostility to peace and truth. Much more than generalized goodwill is needed. What is also needed is a clear-headed and unfrightened analysis of the kind of circle that each system or subsystem is, and the extent to which it will or will not move from coherence dependent on closure to the truth beyond itself.

The beginning of all ecumenical exploration must be the recognition on the part of those taking part of the nature of the circles that are involved. By becoming open to each other and to the world, by the hard and laborious work that I have tried to indicate and describe in this book, they have the chance to move together toward truth. Instead of the circles being impermeably closed, they can lock together in a chain that can better bear the grief and the glory of this world. At the end of *The Mill on the Floss*, George Eliot made a brilliantly perceptive analysis of the dilemmas facing those who live in the closed circles of religion (in this case, Dr Kenn, the Rector of St Oggs, who has gone far to help Maggie Tulliver, but who now must reconcile what the person and what the system demand). First, George Eliot recognized the necessity for casuistry (p 81), despite the bad associations of that word, but then she glimpsed the truth that the purpose of it all is the maintaining of the system provided it sets people free to the love of God and of their neighbour.

> The casuists have become a by-word of reproach; but their perverted spirit of minute discrimination was the shadow of a truth to which eyes and hearts are too often fatally sealed – the truth that moral judgments must remain false and hollow, unless they are checked and enlightened by a perpetual reference to the special circumstances that mark the individual lot.

> All people of broad, strong sense have an instinctive repugnance to the men of maxims, because such people early discern that the mysterious complexity of our life is not to be embraced by maxims, and that to lace ourselves up in formulas of that sort is to repress all the divine promptings and inspirations that spring from growing insight and sympathy. And the man of maxims is the popular representative of the minds that are guided in their moral judgment solely by general rules, thinking that these will lead them to justice by a ready-made patent method, without the trouble of exerting patience, discrimination, impartiality, without any care to assure themselves whether they have the insight that comes from a hardly-earned estimate of temptation, or from a life vivid and intense enough to have created a wide fellow-feeling with all that is human.[46]

A final word: in that entirely new way of seeking union, we need to monitor carefully the style in which we conduct it, because to value the diversity of coherent stories while we move together from coherence to truth carries with it a continuing acceptance of differences. Therefore, the style in which we disagree is so important. Anthony Trollope observed the Church around him and wrote of Mrs Grantly and Mrs Proudie that they were essentially church people, and that consequently it was natural that they should hate each other.

It is a remarkable tribute to Henson and Gore that they never wavered in their affection for each other. In Gore's appeal to the archbishop to stop Henson's consecration, he wrote:

> I need not say with what profound sorrow I have written this protest and appeal. Dr Henson and I have always been friends, and though we have often differed in public, I believe no angry word has ever passed between us or marred our friendship; and I believe him to be personally among the most honourable and courageous of men.[47]

That same evening, Henson wrote to Gore:

> I seize my pen, not to make any comment on your appeal to the Archbishop, but only to reciprocate the affection which you are good enough to express. In these ill times, and in these unkindly circumstances – which we didn't choose, either of us – we mustn't let anything rob us of that.[48]

So yes, we tell much about ourselves and the world in history, in ethics, in science, in aesthetics and in many other ways. What we say is always corrigible and incomplete. But the conducive properties that invade the particular architecture of energy that humans happen to be point to a truth beyond themselves that is not incomplete, the truth to which we give the name of God. God becomes known in us, in the midst, certainly, of our own contingent circumstance. But *what* becomes known is greater than the conducive properties that first signal that truth. When they go, it endures, and we with it, God known beyond contingency in the practice and the presence of prayer – God for us, the means of grace and the hope of glory. And we mustn't let anyone rob us of that.

6

SEX AND SAFETY: A NEW CRISIS FACING RELIGIONS

The strength of religious systems that rely on coherence for the justification, or for that matter the condemnation, of what people say and do, is that they know exactly how to evaluate whatever is said or done. Their weakness is the difficulty they have in accepting question or criticism from outside the boundary of their coherence. Thus it is an entirely legitimate option to establish a strong boundary by living, as do the Old Order Amish, in a style that perpetuates a world before electricity or the internal combustion engine were invented, or by attempting to live, as do some Muslims, in the style that the arRashidun (the first four Caliphs) brought into being. But the coherence that confers legitimacy on systems of this kind requires a wary vigilance in monitoring and maintaining their boundaries.

Those are no doubt extreme examples. Nevertheless, they illustrate why, in general, strongly bounded systems relying on coherence are likely to regard with suspicion the other worlds that surround them. They may of course simply denounce those other worlds, or they may seek to demonstrate from their own legitimizing resources (for example, revelation) the extent to which those other worlds are in fact coherent with their own. That is why, for example, much work in religion and science is devoted to demonstrating (or at least claiming) particular ways in which that reinforcing coherence works - as indeed sometimes it does.

On the other hand, there are those in all religions who seek a somewhat different relationship. It is not one that denies the importance of coherence in the justification of action, speech and thought. However, it accepts (and indeed emphasizes) the importance of communication across the boundaries of particular systems (the secular as much as the religious), a communication which does indeed raise questions and criticisms, but which can also bring encouragement and renewal.

The tension within religious systems is between those who insist on nothing but coherence with the designated source of authority as the necessary mark of belonging truly or legitimately to the system, and those who seek to relate that source of authority to the new worlds in which humans successively live, in a way that allows those new worlds to be a legitimizing part of that which is truly the case – and that is far from saying that 'the new worlds' are always and automatically right in what they say and do. It is nothing like the 'subordinating of revelation to the fads of contemporary fashion', as the accusation is often made.

This tension, created by the handling and evaluation of coherence, produces much of the splendid, though often bewildering, variety of religious ways of life. Living within a coherent system can yield impressive consequences in human life. Yes. But it produces also many of the passionate and intolerant conflicts that divide religious people so bitterly from each other. For those outside those systems, it leads inevitably to a crisis of credibility (amounting often to contempt), because it is impossible to enter into serious conversation, let alone argument, with those who appeal only to coherence within the boundaries of their own system.

It means, therefore, that religions (or religious people) are always in some kind of crisis in relation to each other and to the worlds in which they live. The crises in the past have not been the same. Nevertheless, they have resembled each other in a way that the new crisis facing religions does not. It is the purpose of this final chapter to explore what this new crisis is, how it relates to the arguments in this book, and why it differs from those that have gone before it.

Come back, therefore, to the year 1597. The year is not chosen by chance. It is the year in which Thomas Gresham founded the college in London that bears his name and continues to the present day. Gresham was the Royal Agent in Antwerp who did much to transform London into a commercial centre and who founded the Royal Exchange. He also recognized the importance of 'the new learning' of his time in the development of the fortunes of London, and he founded Gresham College with seven professors, whose task it was (and still is) to give free public lectures to the citizens of London. It was here that the Royal Society was founded. The professorships are in astronomy, divinity, geometry, law, music, physic and rhetoric. To these, commerce was added in 1985.[1]

In 1997, therefore, the College celebrated its 400th anniversary. It is a moment when one can see very clearly in the case of religions a dramatic transition, a transition from a world that Gresham would still have recognized, to one that is increasingly different, from a world in which closed circles of coherence could operate with confidence, to one in which the boundaries of those circles are under a new kind of challenge and threat.

Four hundred years ago, in 1597, the major crisis for religions was one of conflict between religions, or of conflict within religions. Consider what was happening in different parts of the world in that year. It was in 1597 that the Japanese warlord Toyotomi Hideyoshi (1536–98) made his second and more serious invasion of Korea in order to secure a safe passage for his armies as they fulfilled his ambition to conquer China. But on that occasion, as also in an earlier attempt in 1592, it was Buddhist soldier-monks who were at the heart of Korean resistance: they could not stop the massacres and destruction, but as Buddhists they made it clear how religious people, even those committed to non-violence, are capable of going to the limit of resistance, including self-sacrifice, in defending the boundaries of their belief.

Toyotomi Hideyoshi died in 1598, and he was succeeded by Tokugawa Ieyasu (1542–1616). From him the great Tokugawa period in Japanese history, lasting nearly 300 years, takes its name. Tokugawa was the third of the so-called 'three great ones', Oda Nobunaga (1534–82), Toyotomi Hideyoshi and Tokugawa Ieyasu himself, and these three certainly posed a huge crisis for religions in Japan. Toyotomi Hideyoshi had expelled foreign Christians from Japan in 1587 with the words, 'Japan is a country of the *kami*, and for the fathers to come here and preach a devilish law is evil. Since such a thing is intolerable, I am resolved that they shall not stay on Japanese soil. Within twenty days they must return to their own country.' By 1630 a ferocious persecution had destroyed the newly founded Christianity in Japan.

It was not Christianity alone that was in trouble. During this period, Buddhism also lost the great role it had played during earlier times. No longer were Buddhist officials given great honours; no longer did most ordinary people seek their welfare and salvation in the many Buddhist temples scattered throughout the land. As Dumoulin, the great historian of Zen Buddhism, put it:

> Blow by blow, Buddhism suffered painful losses and watched as its position of prominence slipped away. Only the Zen school was able to maintain its special place. ... Buddhism became primarily a popular religion, without any claims of spiritual leadership or of significant influence among the educated classes.[2]

This was a religious crisis indeed, and one which Takuan Soho (1573–1645) recognized and negotiated in such a way that it was Zen Buddhism alone that retained its influence within the new order.

Come a little closer to Gresham's, London. What of India in 1597? In 1597, the Mughal (and therefore Muslim) emperor Akbar was making his great drive to bring the south of India under his control. At first sight, this hardly seems a crisis for religions, because Akbar is famous for his attempt to bring all religions together into his Din-i-Ilahi, his new religious movement of harmony (p 176), and to that end he had established his 'Ibadat-khana', his House of Worship, where people of all religions met to seek common ground in common goodness between the different faiths.

In doing this, was he not simply extending and institutionalizing the unifying vision of Guru Nanak who had died only 58 years earlier, and from whom the Sikh religion is derived? As the later tenth guru, Gobind Singh, was to put this same vision:

> Hindus and Muslims are one.
> The same Reality is the Creator and Preserver of all;
> Allow no distinctions between them.
> The monastery and the mosque are the same;
> So are the Hindu worship and the Muslim prayer.
> Humans are all one![3]

Guru Arjan, the fifth Sikh guru (1563–1606), was guru during the reign of Akbar, and he believed that the Sikh religion was exactly the synthesis that Akbar was seeking. When he completed the gathering together of the hymns of the *Adi Granth*, the Sikh Bible, he included hymns by Hindus and Muslims. Akbar was told that Arjan had gathered hymns attacking Islam and the emperor, so he asked to hear some of them. When he did so, he was so delighted that he cancelled the local taxes.

That hardly seems a time of crisis for religions. And yet, of course, it was, because to some people this inclusive policy was threatening the distinctive truth of their own tradition – in other words, it was threatening the boundaries of the closed circle in which they lived. Among Muslims, the best known of those who felt that way is Ahmad Sirhindi (1524–1624). He had in fact begun his career at the court of Akbar, an experience which had convinced him that religious observance must start at the top. As he put it, 'The ruler is the soul, the people are the body: if the ruler goes astray, the people will surely follow.' But he became certain that this ruler, Akbar, had gone far astray, and he led a vigorous campaign to restore Quran and Sunna to the court and to the people. Almost alone he contested and defeated an interpretation of Islam and of religious experience which was derived from ibn Arabi[4] (whom he called a *kafir*, unbeliever), and which had become widespread among Sufis. In contrast, he believed that it was vital to maintain the boundaries of the closed circle by 'conforming to the *shari'ah* in every detail':

> This rigid adherence in practice was to be a reflection of an equally strict conformity in the spiritual life, and all experiences and 'conditions' were to be examined in the light of the *shari'ah*.[5]

This was for sure a religious crisis of enormous consequence, and it runs right down to the present time: is it necessary for Muslims to live within the closed circle of Quran and Sharia (relying on internal coherence) in order to be *muslim*, or can they integrate insight and understanding from outside the circle, as ibn Arabi did? By calling him a *kafir*, Ahmad was stating that he had ceased to be *muslim*, since a *kafir* is one who is con-demned by God in the Quran as an unbeliever – against whom true Muslims are commanded in the Quran to fight:

> Kill those who do not believe in God, or in the last day, and who do not hold forbidden what God and his Apostle have forbidden, and who do not ac-knowledge the way of truth [din alHaq] from among those who are peoples of the Book [mainly Jews and Christians] until they pay the tax [of recognition of Muslim superiority, *jizya*] from the hand[6] and feel themselves subdued.[7]

The relevance of this to understanding the Taleban and alQaeda is obvi-ous – even though some among the members of those groups (those who engage in indiscriminate violence) cannot be regarded as 'true Muslims'.[8]

And what of Europe? What of that England in which Thomas Gresham was founding his College? To observe that the Spanish Armada had been defeated only nine years earlier will make it obvious how great the crisis was – not just the political crisis, but also that underlying crisis in Christendom of the Reformation. The England of Gresham's lifetime under Edward VI and Mary had swung, on the statute book, between Reform and Rome. Protestants and Catholics had been executing each other, given the chance, with equal determination. English sailors cap-tured at sea were handed over to the Inquisition. 1597 was certainly a year of crisis in religion in Gresham's England.

But in 1597, the fifth book of Richard Hooker's *Of the Laws of Ecclesi-astical Polity* was published, that book of which Izaak Walton wrote that there is in it 'such bowels of love, and such a commixture of that love with reason, as was never exceeded but in holy writ'. Between the extremes of Rome on the one side and of Calvinist Protestants on the other, Hooker sought a middle ground on which people could live in peace and charity with each other. Hooker did not deny that some things are right and others wrong, and he knew that such things must be attended to – indeed, he made clear *how* they should be attended to. But what he resisted was the human tendency to convert what he called outspokenly 'silly things' into matters of mutual hatred – and by that hatred into matters that may destroy the common good:

> These controversies which have lately sprung up for complements, rites and ceremonies of church actions, are in truth for the greatest part such silly things, that very easiness doth make them hard to be disputed of in serious manner.[9]

Hooker made a plea that people would always moderate their own opinion by a discernment of the common good:

> Our wisdom ... must be such as doth not propose to itself το ἴδιον, our own particular, the partial and immoderate desire whereof poisoneth wheresoever it taketh place; but the scope and mark which we are to aim at is το κοινον, the public and common good of all.[10]

Hooker's plea and the Anglican settlement were very extraordinary achievements in a bitter and divided world. The Greek το κοινον refers to that which is held in common, the common good, that which is connected by common origin as in a family. It was a plea not to draw the boundary of the circle of coherence so impermeably that those outside must be rejected – or burned. To observe Anglicans trying to resolve the issue of the consecration of a gay man in an active homosexual relationship (p 2) shows how hard it is for the parties involved not to draw back into extremely closed circles of coherence. Hooker's plea was remarkable and unusual, and had immense consequences for political as well as religious attitudes in England, down to the present day. But it does not in any way diminish the seriousness of the crisis in religion at the time when Thomas Gresham was founding his College.

What, then, is meant by adding the word 'new' to the title – a *new* crisis for religions? Were not the crises for Takuan, for Sirhindi and for Hooker real enough? Indeed they were. But they were all crises *within* religion, within a domain of shared assumptions. Takuan and Tokugawa, Sirhindi and Akbar, Hooker and Catholics and extreme Puritans, were not in dispute about the power and purposes of religion. For sure, Takuan, as a Zen Buddhist, would have been much in dispute with Sirhindi, had he ever met him, about the nature of Allah or of God, but in their respective traditions they were not questioning the worth of religion as they had come to know it, nor, in general, was the society around them. The function of religion, what it does for individuals and for society, was not in question.

Now it is. And that is why the new crisis for religions is of a different kind. In one extensively important area of human life religions are no longer needed as once they were. One of the major reasons why religions have been so important in human life and society for millennia has become unimportant. They are no longer needed as they were. It is a crisis signalled by the words, 'sex and safety'.

Religions and the Protection of Information

To understand what that crisis is, we need to remember that religions have continued so coherently through time because they have been (and still are) highly organized systems. This they have to be because they are protecting and transmitting information that has been (and still is) of the highest possible value to human lives. When we think about religions, we are likely, in the first instance, to think about the great purposes of religion: 'My religion,' wrote Rumi, 'is to live through love.'

> O sudden Resurrection! O boundless, endless, compassion!
> Beyond the sanity of fools is a burning desert
> Where your sun is whirling in every atom; drag me there,
> Beloved, drag me there, let me roast in your perfection.[11]

So, yes, when we think of religions we think of such things as God and prayer, enlightenment and simplicity, sacrifice and sin, mosque, church, temple and synagogue; we think about beliefs and practices which for more than three-quarters of the population of the planet remain, to say the least, important.

There is, of course, a negative list as well, but the fact remains that people living within religious systems can find their darkness transformed into light, and often do. That is why all religions are systems for the coding, protection and transmission of information which has achieved the highest possible value for human communities through the long process of human history. Religions are highly organized information processing systems. They may be much more than that, but that *at least* is what they are. They are organizations to make sure that important information and insight gets passed on from one life to another, and above all, from one generation to another.

So the next question becomes inevitable: what information? Much of it is, of course, that which we regard as characteristically religious, as alluded to in summary above. But part of it has to do with matters that at first sight seem to be so basic that they have little to do with religion as such. It has to do with the fundamental necessities that enable humans to survive and flourish. It is everything that has been endorsed by natural selection and evolution. For how do humans survive and create the next generation? Not by accident but by organization. The worth of a particular form of organization may be tested in many ways, not least by contingent accidents. But the embracing test is that of natural selection. It is natural selection, through the sifting process of evolution, which sets an impartial rule against the experiments of life of whatever kind, on the earth or under the earth, in the air or under the sea. The agents of those experiments or explorations which are best adapted to the conditions all around them survive long enough to replicate more of their genes into another generation; those which are ill adapted may not survive at all.

Looked at from this point of view, bodies have been thought of as gene-survival machines: a chicken is an egg's way of making another egg (Samuel Butler). The genes need protection in order to pass on the code to the next generation: a chicken is the armoured car in which the treasure of the genes is delivered safely to the bank on the other side of town – delivered, that is, into the next generation.[12]

The genes of a chicken are protected twice over, 'belt *and* braces'; in other words, they have two defensive boundaries: the first boundary is the cell inside which the genes are sitting, and the second is the skin: 'Few frontiers hold a world more wondrous in.'[13] The skin is the second defensive boundary of the whole gene-replication process, and that is as true of any human being as it is of a chicken.

But humans have then built a third defensive system outside the boundary of the body: they have built what we call 'culture', so that things like armies, hospitals, traffic lights, schools and microwave meals all play their part in helping humans to survive and flourish. This means that culture is the third defensive skin inside which the gene-replication process sits. So for humans, gene replication is protected, not just by belt and braces, but by a stout piece of string as well.

Religions and Natural Selection

And what has this to do with religion? Everything, because religions are the earliest cultural systems, of which we have evidence, for the protection of gene replication and the nurture of children. Obviously, our early ancestors knew nothing of how gene replication works. But that is irrelevant to the evolutionary point. It is not understanding but successful practice that is measured by survival. That is why religions have always been preoccupied with sex and food, creating food laws and systematic agriculture, and taking control of sexual behaviour, marriage and the status of women.

Religions, or subsystems within religions, are thus highly organized protective systems. It does not mean that all religions are therefore the same simply because they are systems. Obviously not. There are many different styles of organization, ranging from the strongly bounded and hierarchical (such as Vatican Catholicism or Wahhabi Islam) to the weakly bounded but with strong subsystems (such as Anglicanism). At either extreme (and in between), the necessity remains the same, that religions in order to continue through time have to be organized systems in order to secure and transmit the information that human communities have come to value, including the protection of gene replication and the nurture of children.

One early and important reason for religious diversity lies in the fact that there are many different reproductive strategies even in the animal kingdom, let alone in the human. The familiar point is summarized by Vogel:

> All organisms are shaped by natural selection. Since natural selection operates through differential reproduction, this makes reproduction the key phenomenon of evolution. Hence all organisms compete for their own reproductive success which is, in general, the most effective means of maximising personal fitness.[14]

Vogel then went on to make the orthodox point:

> Under certain conditions some individuals of socially living animals may postpone or even forego their own reproduction in order to maximize their inclusive fitness, for instance, by taking the role of 'helpers at the nest', i.e. helping closely related individuals to raise their offspring successfully. Thus, we may find highly sophisticated strategies of transferring as many replicators of 'own' genes to the next generation as possible. Of course animals generally do not consciously engage in strategic actions to pass on their genes, or at least we need not assume that they do. Natural selection, in principle, does favour any behaviour of animals which generates above average reproductive success, as though the actors were consciously seeking a specific goal or result, in this case maximum inclusive fitness.[15]

The plural word in the title of the book, Strategies, makes the point: there are many possible strategies for achieving the rewarded goal of reproduction, and in those strategies the interests of male and female may well be divergent, particularly if the females have to commit a long period of time to gestation, birth and the nurture of the dependent infant. Females frequently have a far longer and costlier commitment to the birth and nurture of the next generation than males. Of course it may be a rewarded strategy for males to protect their mates and the offspring, but on the other hand, it may be a rewarded strategy to take off and seek multiple mates. This is a recipe, if not for 'the battle of the sexes', then certainly for competition between them. The competition of interests and strategies may result in compromise, but it may also result in such extreme measures as infanticide. Langur monkeys, for example, breed in harems, and since there are not enough harems to go round, and since in any case control of a harem is short lived, male monkeys have limited chances of reproductive success. In those circumstances, infanticide on the part of male monkeys is highly adaptive, given the selection pressures on them: if they kill the offspring of their predecessors, they will bring the mothers out of lactation and into oestrus again without too much delay.[16]

What are the best strategies for males and females in the human case? We cannot simply look at the many so-called strategies adopted by other organisms and use them as a template onto which human strategies are mapped as though they are the same (see the discussion of this point on pp 112–13). The difference is obvious: other organisms do not have strategies, humans do. As the sentence quoted above put it: 'Of course, animals generally do not consciously engage in strategic actions to pass on their genes.' The so-called strategies are simply behaviours that have worked and have been rewarded in replication and survival. Humans, however, do engage in conscious and shared strategies that lead to social consequences, including the regulation of mates and the organization of the family and its hierarchy.

It is this that explains why the family is the basic unit of religious organization, even in religions where celibacy is seen as a higher vocation. In almost all religions, the family is far more fundamental than church or temple, synagogue or mosque in the protection and transmission of spiritual information. From this point of view, it is fair to say that one of the greatest of all religious inventions was the family. It is true that Edmund Leach contended in 1967 that 'far from being the basis of the good society, the family, with its narrow privacy and tawdry secrets, is the source of all our discontents.' Described in that way, religions might well agree with him: no one can doubt that the religious fact of families has led to disastrous outcomes, not least where daughters are concerned. But they have also achieved other truths and values in relation to the family, and therefore they have other and better things to say as well.

The point is that there is not one way only in which humans can cooperate in order to maximize the chances of reproductive success. That is why there are so many different forms of family organization. Since success includes bringing up the next generation, it is not surprising that marriage is the commonest strategy of all. But even then, marriage is not a single strategy: there are many different kinds of marriage. For example, marriage may be within a group or class or category (the TV series *Upstairs Downstairs* made that extremely clear), and that is known as endogamy; or it may be outside a group, and that is known as exogamy; or it may be by way of exchange, so that daughters become valuable property; or it may be by capture, after a raid on a nearby village; and so on.

So the strategies are many. But within social groups, it has clearly been an advantage to have stability of expectation. People 'know where they are' and know what is expected of them. The organization of mating and of provision for the nurture of children has reduced conflict and maximized cooperation. There is no one way in which this has to be done. But in whatever way it *has* been done in human history (so far as we have evidence), religions have been the systems which have provided the codes, the sanctions and the endorsements of sexual behaviour, and they have provided also actions and explanations, in ritual and myth, which support the accepted strategy. Each religion tells a story, a great story, into which individuals donate their lives and play their part in turn. The great stories of religion, enacted as well as told, have given unity to a community or to a society.

Religions, Social Strategies and the Status of Women

Religions have thus stabilized social strategies, so that people, in general, agree on the right basic ways to behave and know how to use the vocabularies of approval and disapproval. They are educated (above all in family upbringing) into the direct seeing of conducive properties and into the appropriate emotional and rational responses (appropriateness, that is, within the system). Religions have then given these stable social strategies a continuity through the generations far beyond biology. It is religions which have supplied the maps of approved strategies for reproduction and sexual behaviour in any social group. It is only recently, in many societies in the modern world, that the function of the religious stories in the regulation of sexual behaviour and of marriage has been taken over entirely by secular governments, leading eventually to state control of marriage and birth. The attempts to achieve this in the Western world, while at the same time inheriting from the Enlightenment a view that marriage and sexuality are a private matter in which the state should only intervene minimally, has resulted in the confusion of the present scene.

Not much of this has been good news for women, if good news means having the same status as men in determining the outcomes of their own lives. The status of women has been tied in religions closely to the reproductive cycle, not just the reproductive cycle of the women themselves, but also of the crops and herds on which the group or family depends. Religions endorsed a necessary division of labour which is based on biology, and which therefore paid much attention to menstruation and the availability of women.

The old way of stating this, that women stay at home and men go out to work, is certainly wrong. In many traditional systems, women do a great deal more in terms of work outside, above all in agriculture, foraging and preparation of food. What happens commonly is that women are responsible for birth and the upbringing of children, at least in the early years, and for related activities in the preparation of food, both in the fields and in cooking. Thirty years ago, a photo appeared in *South African Panorama* of an African woman hoeing vegetables. At that date, the journal was trying to persuade the outside world that the separation of husbands and wives in the apartheid system was enjoyed by all. The caption read: 'The women tend the vegetable gardens, not only because their menfolk are away working in nearby Pinetown and Durban, but also because it is an added form of exercise.'

Nevertheless, in traditional systems of that kind, there are clear lines that mark off male from female activities. As the anthropologist Nigel Barley put it of the Dawayo in the Cameroons:

> Within the apparent uniformity of life with the cattle and the fields lies a system of demarcation that would inspire envy in a shipyard worker. Only blacksmiths may forge. Only their women may make pots. Hunters may not keep cattle. Rain-makers and smiths must not meet. Each activity has its responsibilities and potential dangers. Precautions not taken, prohibitions ignored, all have their effects on the community.[17]

Frequently, men relate to a wider environment, in relations with neighbouring villages or eventually states, in hunting, in physical defence and aggression against the outsider, and in organizing the local community; which means that men were far less important from an evolutionary point of view than women. Women indeed are *so* much more important that the male defence of community included the organizing and the control of access to women.

The basic reason is obvious: it is always possible to be sure, at least in a small community, who is the mother of a child, but without strong control it is not so easy to be sure who is the father. It is a reason why polygyny (marrying more than one wife) is far more common than polyandry (marrying more than one husband). In the interviews on which the book *What Muslims Believe*[18] is based, I asked a Muslim why polygamy is allowed to Muslim men but not to Muslim women. He replied:

> The reason is quite simple: you want to know the father of the child. The mother is unmistakeably established in the whole act of procreation. The mother is known. It is the father who would otherwise be uncertain, if a woman married more than one husband.[19]

I then went on to ask him, 'But supposing it were now possible to establish easily ... the genetic paternity of any child, would that open up the possibility that a woman might marry more than one husband?' He answered: 'Oh no, I am afraid that is going beyond our limits.'

'Limits': the word for 'limits' in Arabic, and therefore in the Quran, is *hudud,* or in the singular *hadd.* It means a boundary or a limit set by God, and so it describes the laws laid down by God. Here, as an example, is the Quran on sex and fasting during the fast of Ramadan:

> Permitted to you on the night of the fasts is the approach to your wives. They are your garments and you are their garments. ... So now lie with them and seek what God has prescribed for you, and eat and drink until the white thread [at first light] appears to you distinct from the black thread; then complete the fast till the night appears, and do not lie with them while you are in retreat in the mosques. Those are the limits of God [*tilka hududu 'Llahi*].[20]

And here is the Quran on divorce:

> A divorce is allowed twice; after that, it is a matter of either holding together on equitable terms, or separating with kindness. It is not lawful to take back any of your gifts, except when both parties fear that they would be unable to keep the limits of God. If you fear that they would be unable to keep the limits of God, there is no blame on either of them if she gives something for her freedom. Those are the limits of God, so do not transgress them. If any do transgress the limits of God, those are the wrong-doers.[21]

Hudud: limits, in relation to both food and sex. Here at once, and in miniature, can be seen the powerful importance of religions as organized systems in the domain of gene replication and the nurture of children. Religions established the limits of a life that can evoke approval and the predication of 'good'. They established the codes of behaviour, as well as the sanctions and endorsements to make them stick. And they have worked. The word of God, whether in Bible or Quran or any other scripture, is a very powerful constraint and sanction.

For millennia, therefore, religions have been the social context in which individuals have lived their lives successfully, where success is measured basically in terms of survival and replication, and socially in terms of approval. Success, in this context, is certainly not being measured in terms of individual freedom; and it does not need a feminist to recognize that the strategies adopted by religions to protect gene replication and the nurture of children have usually involved the protection of women and the control of their lives by men.

This does not mean that women cannot have a very high status in religious societies. Frequently the feminine is celebrated in religions as the source of power. Power means that the feminine is not only the obvious source of life and the gift of fertility: she is also the source of death. That is why in India, Mahadevi, the great Goddess, the feminine who becomes manifest in many different guises, is often life-giving in association with a male consort, like Śiva, but death-delivering or death-controlling on her own. As Lalita, for example, she is known as 'horrific' or 'extremely horrific' (ghoratara), yet she is better known as benign and auspicious: she does not derive those benign characteristics from her subservience to her consort, Śiva (as many other manifestations of Goddess do), yet even so, 'Her auspiciousness is made stable and consistent by her relationship with Śiva and therefore, unlike Śiva, she is utterly predictable in her anger and perfectly under control even when it is unclear if she controls herself.'[22] Life and death are the pulse of the feminine, and that is why blood, not least in menstruation, is marked off as both gift and threat.

It is, therefore, wrong to think that in religions women have a subordinate status in all ways. In the home, certainly, the wife and mother is likely to be revered, and that reverence has been translated into worship in many parts of the world. It is equally wrong to think that all women everywhere are seeking to unite against this since they have nothing to lose but their chains. In fact, many women perceive these systems as working well for them also. That is why it is often women who are visible on the streets campaigning for the status quo, campaigning, for example, for the retention of the hijab, or veil, in Islam. Some married women see themselves now, as much as in the past, as having degrees of importance which in their own eyes exceed those of men. In Worlds of Faith, Mrs Pancholi, a Hindu wife, told me very firmly,

> Women are the transmitters of culture in Hindu tradition, and this role lies in the hands of women, and I don't think a man has time, or even the patience, to do that.[23]

This exemplifies the claim made above (p 158) that one of the most important early achievements of religion was the family. In the family, it is possible for women to be, paradoxically, both subordinate and paramount. Women are the transmitters, not simply of life, but also of culture. Where men became important was in building the extended family, because for this men actually had more time than women.

Religions, Society and Sexuality

The extended family in its ordinary sense is important enough, but what religions created were even larger extended families that went far beyond even the kinship group of actual relatives. Religions supplied the metaphors and the rituals through which genetic strangers have been bonded together in a village or in a larger geographical area. In this way, much larger groups have acted together and have cooperated. In the end some religions have dreamed that the whole human race might be a single family, an 'umma as Muslims would call it, a metaphorically extended family, in which, to quote the Christian version of a comparable theme, we are all members one of another (Ephesians 4.25).

It follows that when Margaret Thatcher decided that society does not exist, she was about as far from religious truth and insight as it is possible to get. She said: 'There is no such thing as Society. There are individual men and women, and there are families.'[24] Where there are families, there already is society. How families relate to each other, and how they consti- tute the extended families that religions are, has been differently achieved in different religions, but in all of them there is a wisdom which has been tested and changed through the course of time.

Once the confidence of these larger families is established by their religious validation, then of course even such apparently disadvantageous behaviours as sexual variance can be harnessed - or prohibited: again, there are many different strategies. Take celibacy as an example: this may serve the community, or it may be regarded as aberrant. Celibate Buddhist monks do much for the wider social communities in which they live, even to the extent of taking up arms against aggressors like Toyotomi Hideyoshi (pp 150-1), since they have less to lose. Muslims, on the other hand, regard celibacy as a denial of the purposes of God in creation, even though eunuchs looking after harems were accepted as a diminished risk.

Here, as always, religions produce a bewildering variety of different strategies. But within them all, the resulting religious control has pro- duced high degrees of stability: it has produced moral codes, designations of who may mate with whom (including prohibited relationships), tech- niques and rituals for producing offspring (often of a desired gender), education, protection of women, assurance of paternity by restricting access to women, rules of inheritance and thus of continuity in society.

Eventually, almost all religions (Roman Catholicism being a prominent exception) have made much, in different ways, of the natural distinction between sex and reproduction. Well before the relation between sexual acts and reproduction was better understood, the potential of sex for pleasure and for power was recognized.

This in itself reinforced the male control of women, because promis- cuous or unlicensed sexual activity would clearly subvert that ordering of families in particular and of society in general which was rewarded in natural selection (or, from the limited perspective of the participants, in stable continuity). So whereas male sexual activity outside the reproductive boundary of the family might not be disruptive in terms of reproduction, it clearly would affect the stability of the family as the unit of selection; and in any case, female sexual activity of that kind would certainly be subversive because of the point already made about children (pp 160-1). There is, therefore, a context of religious restriction in relation to sexuality that has been necessary or at least rewarded from the point of view of natural selection and evolution.

Within that context of restriction, the nature of sexuality and sexual feelings has evoked widely differing responses in religions, ranging from a fear of being enslaved to the passions (leading to a dualistic subordination of sexuality, as in Manichaeism or some versions of Christianity) to a delight in sexuality as a proper end in life, as, for example, among Hindus. For Hindus there are four *puruṣārthas*, four legitimate goals in life, and *kama*, which includes delight in sexual pleasure, is one of them.

The same is true elsewhere: almost anywhere where there has not been an inhibiting fear (a rule of reason over passion), the exploration of sexuality has been religiously important. In Eastern religions, in particular, the nature of sexual energy was explored in many directions. Since sexual arousal seems to make its own demands, what might be the consequence if that energy is brought under human control and directed to different ends? In China this lent itself to the quest for immortality and the gaining of strength, in India to the acquisition of power, in different kinds of *puja* or worship, and in tantra. This means that religions recognized early and widely that sex and reproduction are not synonymous: sexual engagement has purposes and pleasures far beyond the limited purpose of the transmission of life.

In all religions, therefore, including Christianity, the union of a man and a woman, transcending the union of male and female in a biological sense, has seemed to be the nearest one can come on earth to the final union with God. The poetry in which this is expressed is extensive, and it transcends the boundaries of religions. On the other hand, it is equally clear in most religions, including Indian, that sex may be an impediment on the way to the final goal. Sex is among those things that may have to be given up if the unqualified love of God is to flourish. This ascetic option gives the highest value to celibacy, chastity and virginity. In Christianity, it became the dominant voice of the official Church, especially in the West. That means it became the voice of men, because only men have control and authority in the Church (since women, until recently, and still in Roman Catholicism and Orthodoxy, cannot be ordained). Thus the subordination of sex and the attempt to make it in effect synonymous either with sin or with reproduction became, within Christianity, a particular strategy through which men kept control and gave to control a new meaning.

The New Crisis

To say, therefore, that religions are concerned with sex and food (p 156) is to say that they are concerned with the protection of gene replication and the nurture of children, and with the exploration of sexuality, and they have done all that very well. *Of course* they have been concerned with much else as well – with all the many consequences, for example, of a much wider somatic exploration and exegesis (p 49). It is precisely the immense importance of all that has been achieved in religions which reinforces the reason why religions are protective systems: they protect not only gene replication, but also virtually everything else that has been indispensable for human life and flourishing. It is all far too important to be left to chance. It is everything from sex to salvation. It is, therefore, information that has to be organized if it is going to be saved and shared and transmitted. Religions are systems to do exactly that. And while we may think of this information primarily in terms of items which seem to us more obviously religious, in terms, let us say, of gurus or of God, in fact the fundamental information concerning sex, family and food is equally important: without that, at least in earlier times, the rest could never have got off the ground and have been protected and transmitted so success- fully.

And now, at last, it becomes obvious what this 'new crisis' is that is threatening or challenging religions at the present time. It is that religions no longer seem, in many societies, necessary for the protection of gene replication and the nurture of children. Think of what has happened over the last 100 years, at least in technological and affluent societies: the rates of infant mortality have dropped so that we no longer need the insurance of multiple births; techniques of contraception have been developed which reduce the risk of unwanted pregnancy, as much outside marriage as within; smaller families and the better control of when children are conceived have contributed to the emancipation of women from the obligation to be available for reproduction. In many societies, what was once an all-important function of religions, to protect and to enhance the probability of gene replication, has disappeared. Religions are no longer needed as protective systems for this purpose. The pervasive control of religions in the fundamental domain of sex and food may have worked well for millennia – indeed, it has worked well, since otherwise none of us would be here. But it is no longer needed from the point of view of natu- ral selection, with the clear caveat that the new styles of sexuality and reproduction have themselves to be tested from that point of view: it is far too early to know what the impact of AIDS on styles of sexuality is going to be. However, at the moment, religions are no longer necessary to secure the goal of gene replication.

What do religions do in this new circumstance? They adopt, of course, many different strategies, of which one is obvious: they do nothing – or rather, they carry on as usual. That is the crisis facing religions. Because religions have become through time such highly organized and effective systems, in which sex and reproduction are integrated into a coherent system, it seems immensely threatening if sex is pulled out of the system. This can be seen most obviously in the Roman Catholic, or more accurately Vatican Catholic, insistence that the unitive and the procreative functions of sex cannot be divorced. *Humanae Vitae* insists that 'each and every marriage act must remain open to the transmission of life',[25] and that 'every action which, whether in anticipation of the conjugal act, or in its accomplishment, or in the development of its natural consequences, proposes, whether as an end or as a means, to render procreation impossible' is intrinsically evil.[26]

This is based on an appeal to natural law (see pp 79–82) and to the right of the Church to pronounce on such matters (§4). Natural law is not of course to be confused with that which happens naturally: 'The natural law,' says the *Catholic Catechism*, 'expresses the original moral sense which enables man to discern by reason the good and the evil, the truth and the lie.'[27] But then it is simply a matter of rhetoric to claim that the unitive and procreative functions, sex and reproduction, cannot be separated. In most religions they *are* separated, so it is simply not true that there is a natural moral discernment that they cannot be separated. In nature, in any case, they certainly can be separated, even in the most obvious sense that the words 'sex' and 'reproduction' are not synonymous. Certainly all organisms have to reproduce if genetic survival is to be ensured, but they can do this by asexual as well as sexual means. Asexual reproduction of single-cell species, such as protists or blue-green algae, is comparatively simple: it involves duplication of chromosomes followed by a division. Sexual reproduction is vastly more complicated, and far more costly, as a behaviour.

So what are the evolutionary advantages of sexual reproduction? Part of the answer lies in the way it increases genetic diversity. But part of the answer lies in the fact that sexual activity serves more purposes naturally than reproduction alone. In fact, one of the major rewards of unitive sex being divorced from reproductive sex, and not just in the infertile periods, lies in the bonding and continued commitment of each to other, above all of male to female, though it could equally be male to male or female to female. The protective advantages are obvious, especially for women: men do not simply seek to replicate their genes with the maximum number of partners. They remain committed to the investment in a single partner. To put it as simply as possible (and this is almost exactly the opposite of what *Humanae Vitae* maintains), there is, in the human case, far more to sex than reproduction. It is essential and natural, in the human case, that not all sexual acts should be open to the transmission of life. Most people, as they grow older, come to realize that there is more to life than sex. What human and religious experience teaches us is that there is more to sex than life.

The Vatican claims can of course be justified within its own closed circle of coherence. That is not the issue, or the new crisis. The real issue is whether it is necessary for religions to defend without change systems which have worked so well for so long, when circumstances have changed in a radically disjunctive way. That is the issue of the closed circle. To what extent, if at all, can knowledge or insight be allowed from the out-side into a circle of internal coherence if it challenges a proposition which is justified within the circle? If one part goes, is not the whole threatened, particularly in a system in which matters of faith and morals can be de-fined infallibly?

Not that Vatican Catholicism is alone in this. Any religion that relies on inerrant revelation will be comparable. Thus in Islam, whatever is allowed or forbidden in the Quran is absolute. There is much in human life and behaviour that is, by the mercy of God, left open, but the Quran is not open to change or negotiation. From the Quran and from the example of the Prophet and his Companions, it is clear what the purpose and practice of marriage must be: the primary purpose of marriage is the service and worship (*'ibadah*) of God, and that is achieved by living to-gether as God wills. Only then is the second purpose, the birth of children, brought into context. What does God will? Doi, summarizing Islamic law, makes this clear:

> The man, with his aggression, is charged with what is called the 'instrumental' functions: maintenance, protection, dealings with the outworldly matters and leadership within the family. The woman is entrusted with caring for and rear-ing the children, organising the home, and creating the loving atmosphere inside her matrimonial home. Work or trade are not prohibited to woman in Shari'ah provided they do it within the framework of modesty and with the permission of the husband; they are not recommended to undertake such ac-tivities unless there is a justification for them to go to work and should be without prejudice to their husband's rights.[28]

This is an example of religion as a protective system, of the kind so well rewarded by natural selection. Can this change without calling the Quran into question? The crisis is the same. Of course the Vatican and Sunni Islam do not agree on all matters. In fact, in Islam contraception *is* permitted for valid reasons, and those reasons are often listed: the most important are those that have to do with the health and well being of the mother or of existing children. Where the Vatican and Islam *are* agreed is in defending the status of religion as the protective system in which alone sexual activity and gene replication should occur, and must occur, in the ways they say.

The new crisis for religions is, therefore, to know whether they need for their own survival to maintain the same systems of control over gene replication that have served them so well for so long. Is it the only inevitable policy for religions to reiterate their control of reproduction and sexual behaviour? The disaster of doing that is obvious: gene replication no longer requires, in many parts of the world, the protection that religions used to supply, and sex and reproduction are increasingly being separated. For religions to insist that this is wrong, on grounds that can be justified only within their own closed circles of coherence, is to drive a schism into the human community.

The schism may not immediately show, because religions, as a consequence of all their other myriad explorations and discoveries, offer so much more than their aboriginal protection of gene replication that many people will adhere to the religion in general and abandon its imperatives on sex and reproduction. But the incoherence, or for that matter the hypocrisy, is dangerous: it means that religions get identified with a recalcitrant defence of the indefensible, or at least of the unnecessary, and that consequently the wisdom that they have acquired on other matters gets lost. Islam and the Vatican came into an unlikely alliance at the Rio and Cairo Conferences on matters of world population. The effect of this was to make far more extreme the positions taken by others at the UN conference on women at Beijing in 1995. The Vatican was far from being alone in opposing some of the proposals that were made there: an international right to abortion, for example, or the use of abortion as a means of family planning, or the proposal that universal human rights are not universal. But the Vatican made the opposition to such proposals more difficult by insisting so unequivocally on a system of protection and control in relation to sex and reproduction which no longer serves its original purpose. It seemed to be suggesting that only those who agree that contraception is intrinsically evil have a moral right to speak.

And then what happened? It was to women at the Beijing conference that Pope John Paul II addressed his 'Letter to Women' (10 July 1995), in which, while he apologized for the objective wrongs done to women by 'not just a few members of the Church', he then promptly went on to perpetuate them by insisting that the genius of women is of such a kind that it cannot be exercised in ministerial priesthood.

The tragedy of all this is that the wisdom and experience of religions on other matters gets lost. If they are no longer necessary for the protection of gene replication, and yet they insist that they are, they risk becoming incredible on other matters as well. What about the nurture of children? What have religions learned here that might still be of value? And what of the other end of life? What of senescence, of growing old? Both nurture and ageing have evolutionary advantages, although they carry with them high costs. Here, exactly as with the case of gene replication, religions in the past have exploited the necessities and the advantages of evolutionary constraints, and they have made out of them something transcendently human. Religions have so much to say, from well-winnowed experience, about the values of old age; they have even more to say about the value of death.[29] They deal in so many vital ways with what Aristotle called ἐυδαιμονια, human flourishing.

But are they worth hearing? The new crisis for religions is that if they defend a system which is no longer needed for its original purpose, the protection of gene replication, they will seem to be no longer needed for any of their other purposes, including salvation and enlightenment, including also the nurture of children, the attainment of wisdom, the values of age and the goals of life.

So this new crisis for religions is simple to see, far less simple to solve. Religions are the systems that have controlled gene replication and the nurture of children for many thousands of years. They have done this so well that they have also been the context in which the great discoveries and achievements of human enterprise have been secured and have been passed on as opportunity from one generation to another.

Most of this remains as true now as it has been in the past: the opportunities of religion, to create the greatest goodness and beauty in mind and spirit and behaviour, to find God by being found by God, to grasp the nettle and to grasp one's neighbour as being not other than oneself, none of this has disappeared. What has disappeared is the necessity for religions to guard and protect the process of gene replication. And the more a religion identifies itself with that necessity and refuses to relinquish it, the more absurd it becomes. The more it insists that its old protection of reproductive activity belongs to the intrinsic essence of its truth, the less people will care to listen to it on all those other matters, those opportunities for the transfiguration of human life.

That is what it means to say that religions, when they do this, drive a schism into the human community: they diminish our human possibility. Religions have acquired so much truth and so much wisdom through the course of time that they should be way out in front showing how to live in this new world in ways that seek what Hooker called 'the common good'. That must include in our time accepting and affirming with gratitude the emancipation of women, not from religion, but from the now unnecessary restrictions and protection which religions used to exist to provide.

Will they give the lead in this way? One kind of answer is given by the ways in which people live their lives. An authoritative answer can be given only by those who write the catechisms and the handbooks of sharia, the responsa and the applications of dharma. At the moment it looks as though it will not happen, and the human loss will be great:

> For I have seen the ways that lead away
> Beyond the night, and on to endless day:
> Will you, my friend, step with me, break this bread,
> Or stay in safety, safe among the dead?

NOTES

Introduction (notes to pages ix–x)

1   C. Ricks (ed), *The Poems* of *Tennyson* (London: Longmans, 1969), p 1035.

2   C. Woodham-Smith, *The Reason Why* (London: Constable, 1953).

3   R. Dawkins, interview in the *Daily Telegraph*, 31 August 1992, p 11. The interviewer, Mick Brown, not surprisingly commented, 'But that, of course, is precisely the question which people do continue to ask.' 'They may continue to do what they like,' Dawkins says briskly. 'That's their problem. It doesn't mean it's a legitimate question to ask.'

4   C. Darwin, *The Origin of Species* (New York: New American Library, 1958) ch. 14, p 406. See also the list of 'why' questions when he considered the 'complex and singular rules' governing the sterility of first crosses and of hybrids, ch. 9, p 266. When he grouped difficulties with his theory under four heads, one set was under 'why' questions, another under 'how', ch. 6, p 158.

5   Quoted in J. Gleick, *Genius: Richard Feynman and Modern Physics* (London: Little, Brown, 1992), p 357. Even more to the point, Feynman's career as a physicist took off in a Nobel-prize direction within a week of his arrival at Cornell when he saw someone in the cafeteria throw a plate in the air and wondered *why* the medallion on the plate rotated twice as fast as the plate wobbled: 'I had nothing to do, so I start to figure out the motion of the revolving plate. I discover that when the angle is very slight, the medallion rotates twice as fast as the wobble rate – two to one. It came out of a complicated equation! Then I thought, "Is there some way I can see in a more fundamental way, by looking at the forces or the dynamics, why it's two to one?" I don't remember how I did it, but I ultimately worked out what the motion of the mass particles is, and how all the accelerations balance to make it come out two to one. I still remember going to Hans Bethe [a theoretical physicist] and saying, "Hey, Hans! I noticed something interesting. Here the plate goes around so, and the reason it's two to one is ..." and I showed him the accelerations. He says, "Feynman, that's pretty interesting, but what's the importance of it? Why are you doing it?" "Hah!" I say. "There's no importance whatsoever. I'm just doing it for the fun of it." His reaction didn't discourage me; I had made up my mind I was going to enjoy physics and do whatever I liked. I went on to work out equations of wobbles. Then I thought about how electron orbits start to move in relativity. Then there's the Dirac equation in electrodynamics. And then quantum electrodynamics.

Notes to pages xi–4

And before I knew it (it was a very short time) I was "playing" – working, really – with the same old problem that I loved so much, that I had stopped working on when I went to Los Alamos [to work on the atom bomb]: my thesis-type problems; all those old-fashioned, wonderful things. It was effortless. It was easy to play with these things. It was like uncorking a bottle: Everything flowed out effortlessly. I almost tried to resist it! There was no importance to what I was doing, but ultimately there was. The diagrams and the whole business that I got the Nobel Prize for came from that piddling around with the wobbling plate' (R.P. Feynman, *"Surely You're Joking, Mr. Feynman!" Adventures of a Curious Character*, ed E. Hutchings (Toronto: Bantam Books, 1986), pp 157f).

6   Epilogue, lines 17–20, in Ricks (ed), *The Poems of Tennyson*, p 1308.

7   *New Scientist*, 9 February 1984, p 34.

8   For a summary, see J.W. Bowker, *Is God a Virus? Genes, Culture and Religion* (London: SPCK, 1995), pp 28–35.

Chapter 1. Conflict and the Reasons Why

1   J.G. Lockhart, *Cosmo Gordon Lang* (London: Hodder & Stoughton, 1949), p 290.

2   Letter to Mrs David Ogilvy, 25 July 1851.

3   See J.W. Bowker, 'The Religious Understanding of Human Rights and Racism', in D.D. Honoré (ed), *Trevor Huddleston: Essays on His Life and Work* (Oxford: Oxford University Press, 1988), p 154.

4   According to the *Guardian*, 28 October 1984, Jenkins said during a radio discussion: 'After all, a conjuring trick with bones only proves that somebody's clever at a conjuring trick with bones.'

5   See, for example, J.W. Bowker, *Licensed Insanities: Religions and Belief in God in the Contemporary World* (London: DLT, 1987); and *Is God a Virus?*, Part II, 'Why Are Religions So Dangerous?'

6   See R. Rucker, *Infinity and the Mind: The Science and Philosophy of the Infinite* (London: Paladin, 1984), pp 4–9.

7   The threat of secularism is not that it is an alternative 'ism', a thought-out and coherent system, although it is often talked and written about as though it is. The threat to some religious leaders and people of what is called 'secularism' lies in its commitment to what I have called 'the preferential option for options'. On this and on 'the myth of secularisation' see my article in *The Oxford Dictionary of World Religions* (Oxford: Oxford University Press, 1997), pp 871f.

8   Butler wrote in full: 'When I say I can make you understand "why" this is so, I only mean that I can answer the first "why" that anyone is likely to ask about it, and perhaps a why or two behind this. Then I must stop. This is all that is ever meant by those who say that they can say "why" a thing is so and so. No one professes to be able to reach back to the last "why" which anyone can ask and to answer it.

Notes to pages 5-7

Fortunately for philosophers people generally become fatigued after they have heard the answer to two or three "whys" and are glad enough to let the matter drop' (G. Keynes and B. Hill (eds), *Samuel Butler's Notebooks* (London: Jonathan Cape, 1951), p 134).

9   P. Coveney and R. Highfield, *Frontiers of Complexity: The Search for Order in a Chaotic World* (London: Faber & Faber, 1996), pp 328f.

10   D. Johnston, *Religion: The Missing Dimension of Statecraft* (Oxford: Oxford University Press, 1994); (ed), *Faith-based Diplomacy: Trumping Realpolitik* (Oxford: Oxford University Press, 2003). For an updated summary of his arguments, see his 'Faith-based Diplomacy: Trumping Realpolitik', in *Interreligious Insight*, II (2004), pp 72-7.

11   Johnston, *Religion: The Missing Dimension of Statecraft*, p 13.

12   J.W. Bowker, 'Only Connect ...', *Christian*, VII (1982), p 66.

13   On the importance and application of the frame problem, see my *Is God a Virus?*, pp 54, 252-4.

14   J.W. Bowker, 'God, Spiritual Formation, and Downward Causation', *Theology*, CVII (2004), p 83, quoting *Is God a Virus?*, p 104.

15   R.W. Miller, *Fact and Method: Explanation, Confirmation and Reality in the Natural and the Social Studies* (Princeton: Princeton University Press, 1987), p 86.

16   For an excellent account of the meaning and possibility of special divine action and general divine action, see N. Saunders, *Divine Action and Modern Science* (Cambridge: Cambridge University Press, 2002).

17   P. Levi, *If This is a Man* (London: Penguin, 1979), p 35. In 'Science and Religion: Contest or Confirmation?' in F. Watts (ed), *Science Meets Faith* (London: SPCK, 1998), pp 95-119, I have drawn out the importance of the question 'Why?' for Wagner in resisting a Dawkins-type scientism in the nineteenth century: 'Wagner did not deny that science has a limited truth to tell, because it is a consequence of what he called, in "The Music of the Future", "the ineradicable quality of the human perceptive process, which impelled man to the discovery of the laws of causality, and because of which he involuntarily asks himself, in the face of every impressive phenomenon - "Why is this?" Far from disappearing into an abstraction of music from the real world in order to produce pure beauty, Wagner argued that the total work of art returns the spectator *to* the real world, and offers much richer answers to the question, "Why?":

Notes to pages 7–12

The drama, at the moment of its realistic, scenic presentation awakens in the spectator real participation in the action presented to him; and this is so faithfully imitated from real life (or at least from the possibilities of it), that the sympathetic human feeling passes through such participation into a state of ecstasy which forgets that momentous question "Why?" and willingly yields itself up to the guidance of those new laws through which music makes itself so strangely intelligible and at the same time – in the deepest sense – gives the only correct answer to that "Why?"

So there are other and, humanly speaking, far more important questions than physics can possibly answer. What questions? Wagner's operas are the answer." I then go on to consider *Parsifal* and its exploration of suffering and sin as an example.

18 L. Sterne, *The Life and Opinions of Tristram Shandy* (London, 1948), p 77.

19 Rollin Chamberlin of the University of Chicago wrote: 'Wegener's hypothesis in general is of the footloose type, in that it takes considerable liberty with our globe, and is less bound by restrictions or tied down by awkward, ugly facts than most of its rival theories.' See P. Moore, $E=mc^2$ (London: Quintet, 2002), p 77.

20 G.R. Elton, *Studies in Tudor and Stuart Politics and Government* (Cambridge: Cambridge University Press, 1983), vol 3, p 186.

21 M.H. and R. Dodds, *The Pilgrimage of Grace 1536–1537 and the Exeter Conspiracy, 1538* (Cambridge: Cambridge University Press, 1915).

22 See the review in M. Bowker, 'Lincolnshire 1536: Heresy, Schism or Religious Discontent?', in D. Baker (ed), *Studies in Church History: Schism, Heresy and Religious Protest* (Cambridge: Cambridge University Press, 1972), pp 195–212.

23 Galatians, 2.20.

24 Bhagavadgita, 18.66.

25 Quran, 65.3.

26 'Concursive writing', in relation to the Bible and debates about inspiration, is explained in my *Is God a Virus*, pp 260f.

27 F. Maclean, *A Person from England and Other Travellers* (London: Jonathan Cape, 1958), pp 21f. See also H.P. Palmer, *Joseph Wolff: His Romantic Life and Travels* (London: Heath Cranton, 1935), p 198.

28 John Paul II, *Crossing the Threshold of Hope* (London: Jonathan Cape, 1994), pp 54f., 53.

29 J.D. Caputo (ed), *The Religious* (Oxford: Blackwell, 2002), p 3.

30 F. Jacob, *The Possible and the Actual* (Seattle: University of Washington Press, 1982), p 68.

Chapter 2. The Appeal to History

1 Braley (ed), *Letters of Herbert Hensley Henson* (London: SPCK), p 97.

Notes to pages 12–17

2 Mussolini began to plan the invasion of Ethiopia/Abyssinia in 1928, partly to avenge the Italian defeat in 1896, and also to insist on the Italian interpretation of the Treaty of Uccialli of 1889. According to their own interpretation, Ethiopia was held to be a protectorate of Italy. That is why the Italians claimed that they could invade Ethiopia, as they did in 1935, without declaring war. They occupied most of the country until it was liberated by British, Free French and Ethiopian troops in 1941.

3 Henson wrote a long autobiography called *Retrospect of an Unimportant Life*. References here are to the one-volume edition combining volumes 1 and 2 with separate pagination (Oxford: Oxford University Press, 1942–43).

4 *Ibid.*, vol 2, p 411.

5 Lord Halifax had been appointed foreign secretary on 25 February 1938. He was far from sharing Henson's opinions. According to Lord Boothby, 'When Halifax went to see the Nazi leaders in Germany, Lloyd George said it was like sending a curate to visit a tiger; he wouldn't know whether it was growling in anger or fun; and in either case he wouldn't know how to reply. So it proved' (R.J.G. Boothby, *Recollections of a Rebel* (London: Hutchinson, 1978), p 147).

6 Henson, *Retrospect*, vol 2, p 410.

7 *Ibid.*, vol 1, p 91.

8 G.V. Routh, *Towards the Twentieth Century* (Cambridge: Cambridge University Press, 1937), pp 126f.

9 Braley (ed), *Letters of Herbert Hensley Henson*, p 79.

10 William Temple (1881–1944) was the rector of St James's, Piccadilly. He became archbishop of York in 1929, and of Canterbury in 1942.

11 Charles Gore (1853–1932) was a leading, though also critical, advocate of High Church principles in the Church of England. He had become bishop of Oxford in 1911.

12 Henson, *Retrospect*, vol 1, p 207.

13 *Ibid.*, p 208.

14 *Ibid.*

15 Braley (ed), *Letters of Herbert Hensley Henson*, p 209.

16 In G.L. Prestige, *The Life of Charles Gore* (London: Heinemann, 1935), the chapter covering this same year, 1917, is entitled, 'The Last Straw'.

17 Henson, *Retrospect*, vol 1, p 213.

18 Michael Ramsey (1904–88) was at the time a canon-professor at Durham, and became archbishop of Canterbury in 1961.

19 Braley (ed), *Letters of Herbert Hensley Henson*, p 169.

20 W. Sanday, *The Life of Christ in Recent Research* (Oxford: Clarendon Press, 1907), p 139: 'That is really the meaning of all Christian theology. The facts come first; the formulae, or groups of formulae, which express and partially explain the facts by correlating them with the whole body of belief, come afterwards.'

[21] *Ibid.*, p 37.

[22] Sanday described the painstaking nature of the work he envisaged: 'When I came back to Oxford as Ireland Professor four and twenty years ago, the doctrine that I ventured to preach was: Don't let us be too ambitious; let us plan our work on a large scale, and be content to take the humbler departments first. Let us make sure of our ground as we go on. Let us begin by seeing that we have trustworthy texts; then let us take up literary problems, and work them out as well as we can; let us practise our hands on commentaries and the like. In this way we shall gain experience, and make ourselves fit to aim at higher things' (*Ibid.*, p 38).

[23] Hirata Atsutane, *Kodo Taii*, in *Hirata Atsutane Zenshu*, I, pp 22f.; quoted in Ryusaku Tsunoda et al (eds), *Sources of Japanese Tradition* (New York: Columbia University Press, 1964), vol 2, p 39.

[24] L. Hanke, *Aristotle and the American Indians: A Study in Race Prejudice in the Modern World* (Chicago: Henry Regnery, 1959), pp 47, 49.

[25] Demosthenes, denouncing Meidias, in S.H. Butcher (ed), *Demosthenis Orationes* (Oxford: Clarendon Press, 1907), vol 2.1, §150. On the development of the concept of 'the barbarian', see E. Hall, *Inventing the Barbarian* (Oxford: Oxford University Press, 1991).

[26] Hirata Atsutane, in Ryusaku Tsunoda et al (eds), *Sources of Japanese Tradition*, vol 2, p 41.

[27] Akbar (1542–1605) was one of the ablest rulers of Mughal India. To quote from my entry in *The Oxford Dictionary of World Religions*: 'During the latter part of his reign, Akbar, while maintaining that he remained Muslim, promulgated Din-i-Ilahi (Divine Faith, also called Tawhid-i Ilahi) as a new religion for his empire. It was a syncretization of various creeds and an attempt to create a pure theism. Although he was illiterate himself, he founded an Ibadat-khana (house of worship) where leaders of different religions could discuss their faiths. It appears that Akbar was genuinely interested in reconciling religions: "I try to take good from all opinion with the sole object of ascertaining the truth!" However, Akbar's Din-i-Ilahi met with very little success (it was strongly opposed by Ahmad Sirhindi), and it died with him.'

[28] 'Akbar's Dream', in Ricks (ed), *The Poems of Tennyson*, pp 1,446f., lines 113–18.

[29] *Ibid.*, lines 182–6.

[30] D.W.Y. Kwok, *Scientism in Chinese Thought, 1900–1950* (New Haven: Yale University Press, 1965), p 113.

[31] Wu Zhi-hui, quoted in *ibid.*, p 49.

[32] B. Franklin, 'The Savages of North America', in J. Bigelow (ed), *Complete Works* (New York, 1887–88), vol 9, p 25.

Notes to pages 20–23

[33]   O. Goldsmith, *The Citizen of the World, or Letters from a Chinese Philosopher Residing in London to his Friends in the Country* (London: Cooke, 1799), 2 vols, p 12.

[34]   T. Reid, *The Works of Thomas Reid, D.D.* (Edinburgh, 1812), vol 1, p v.

[35]   S. Hawking, *The Universe in a Nutshell* (London: Bantam Press, 2001), p 21.

[36]   G. Gamow, *My World Line: An Informal Biography* (New York: Viking Press, 1970), p 150.

[37]   The power of the 'grand narratives' (so mistrusted by Henson) is even more dramatically illustrated in the way in which Arthur Eddington 'confirmed' Einstein's theory in 1919 on the basis of an expedition to an island off the coast of West Africa to make observations during an eclipse of the sun. The claim was made that measurements of the bending of starlight by the sun agreed with the theoretical value calculated by Einstein. The result was presented by distinguished scientists (for example, J.J. Thomson, President of the Royal Society) as an immense endorsement of Einstein's theory. In fact, it is now known that Eddington's observations and photos were greatly impeded by cloud cover. His team obtained 16 plates, of which only five were usable – and they did not agree with the results obtained by other observers. Because Eddington accepted the 'grand narrative', he had no hesitation in cutting corners. For this episode, see J. Earman and C. Gilmour, 'Relativity and Eclipses: The British Eclipse Expeditions and Their Predecessors', *Historical Studies in the Physical Sciences*, XI (1980), pp 49–85; H. Collins and T. Pinch, *The Golem: What You Should Know about Science* (Cambridge: Cambridge University Press, 1998); and J. Waller, *Fabulous Science: Fact and Fiction in the History of Scientific Discovery* (Oxford: Oxford University Press, 2002), pp 48–63.

[38]   In *The Prof: A Personal Memoir of Lord Cherwell* (London, Macmillan, 1959), Roy Harrod gives a graphic, eyewitness description of this failure of understanding, in his account of Lindemann's debate at Oxford with two prominent philosophers, J.A. Smith and H.W.B. Joseph.

[39]   On domain assumptions, see Chapter 4, note 65.

[40]   T.S. Eliot, *Four Quartets* (London: Faber & Faber, 1944), pp 37, 43.

[41]   *The Times*, 8 June 2002, p 27. On another occasion, T.S. Eliot refused to unpack the meaning of what he had written. See my 'Religious Studies and the Languages of Religions', *Religious Studies*, XVII (1981), p 431, where Dylan Thomas's comparable response to Edith Sitwell is also recorded.

[42]   'Whatever the fate of leading families may be, the republic endures for ever' (*Annals*, 3.6).

[43]   R. Lowie, *History of Ethnological Theory* (New York: Farrer & Rinehart, 1937).

[44]   See, p 17.

[45]   Lowie, *History of Ethnological Theory*, p 25.

[46]   *Ibid.*, p 25.

Notes to pages 24–28

[47] R.A. Shweder and R.A. LeVine, *Culture Theory: Essays on Mind, Self and Emotion* (Cambridge: Cambridge University Press, 1984), p 195.

[48] J.D. Schomberg, *The Theocratic Philosophy of History* (London: Whittaker and Co, 1839).

[49] A.L. Kroeber, 'The Eighteen Professions', *American Anthropologist*, XVII (1915), pp 287.

[50] Kroeber, *California Kinship Systems*, University of California Publications in American Archaeology and Ethology, XII (1917), pp 339–96.

[51] 'Memes' have been proposed as cultural items that replicate along lines analogous to genes, thus explaining cultural diversity, or what Dawkins called 'the immense differences between human cultures around the world, from the utter selfishness of the Ik of Uganda, as described by Colin Turnbull, to the gentle altruism of Margaret Mead's Arapesh' (*The Selfish Gene* (Oxford: Oxford University Press, 1976, 1989), p 191). The analogy from gene to meme is far too weak to sustain the explanation. For a criticism, see my *Is God a Virus?*, pp 68–72.

[52] W.J. McGee, 'Some Principles of Nomenclature', *American Anthropologist*, VIII (1895), p 281.

[53] L.F. Ward, review of G.H. Scribner, *Where Did Life Begin?...*, in *American Anthropologist*, VI (1904), p 152. The theory was that the earth must originally have been too hot to sustain life, so that life must have begun at the two poles.

[54] F. Boas, 'Rudolph Virchow's Anthropological Work', *Science*, XVI (1902), p 443.

[55] See R.W. Miller, *Fact and Method: Explanation, Confirmation and Reality in the Natural and the Social Studies* (Princeton: Princeton University Press, 1987), pp 15–59, and especially pp 28f.: 'In addition to discouraging certain investigations, the covering-law model has encouraged research programs emphasizing the discovery of general laws in the social sciences. ... Commitment to the covering-law model has kept such goals alive in the face of continual disappointment.'

[56] Boas, *Race, Language and Culture* (Chicago: University of Chicago Press, 1982), p 275; originally 'The Limitations of the Comparative Method of Anthropology', 1896.

[57] *Ibid.*, p 276.

[58] *Ibid.*, p 642.

[59] For a discussion of the nomothetic ambition, see my *The Sense of God: Sociological, Anthropological and Psychological Approaches to the Origin of the Sense of God* (Oxford: Oneworld, 2nd edition, 1995), ch.1.

[60] Boas, *Race, Language and Culture*, p 637.

[61] G. Stocking, 'From Physics to Ethnology: Franz Boas' Arctic Expedition as a Problem in the Historiography of the Behavioral Sciences', *Journal of the History of the Behavioral Sciences*, I (1965), pp 53–66.

Notes to pages 29–35

62 Boas, *Race, Language and Culture*, p 279.

63 Kroeber, 'History and Science in Anthropology', *American Anthropologist*, XXXVII (1935), p 542.

64 Kroeber did not go so far as to deny the dimension of time, but made it relative, on the grounds that if 'the essential quality of the historical approach is an integration of phenomena, and therefore ultimately an integration in term of the totality of phenomena, it is obvious that the time relations ... enter into the task. I am not belittling the time factor; I am only taking the stand that it is not the most essential criterion of the historic approach. Space relations can and sometimes must take its place' (Kroeber, *The Nature of Culture* (Chicago: University of Chicago Press, 1952), reprinted in T. Parsons et al, *Theories of Society: Foundations of Modern Sociological Theory* (New York: The Free Press, 1965), p 547).

65 Kroeber, in *ibid.*, p 1,033.

66 Kroeber, 'The Superorganic', *American Anthropologist*, XIX (1917), p 199.

67 E. Sapir, 'Do We Need a "Superorganic"?', *American Anthropologist*, XIX (1917), pp 442–3.

68 Sapir, 'The Status of Linguistics as a Science', *Language*, V (1929), pp 207–14, reprinted in D.G. Mandelbaum (ed), *Selected Writings of Edward Sapir on Language, Culture, and Personality* (Berkeley: University of California Press, 1949), p 162.

69 *Ibid.*, p 209.

70 B.L. Whorf, 'An American Indian Model of the Universe', in R.M. Gale (ed), *The Philosophy of Time* (Sussex: Harvester Press, 1978), pp 378f.

71 *Ibid.*, p 370.

72 *Ibid.*, p 386.

73 J.A.M. Meerloo, 'The Time Sense in Psychiatry', in J.T. Fraser (ed), *The Voices of Time* (London: Allen Lane, 1968), p 644.

74 C. Lévi-Strauss, *The Savage Mind* (Chicago: Chicago University Press, 1967), pp 233f.

75 *Ibid.*, p 234.

76 H.W. Walsh, *An Introduction to the Philosophy of History* (London: Hutchinson, 1951), p 59.

77 See H.W. Walsh, 'The Politics of Historical Interpretation', in *The Tropics of Discourse* (Baltimore: Johns Hopkins University Press, 1978), p 70.

78 B. Croce, *History: Its Theory and Practice*, trans D. Ainslee (New York: Harcourt Brace, 1923), p 233.

79 Croce, *A Second Collection* (Philadelphia: Westminster Press, 1974), p 71.

80 J. Derrida, 'Structure, Sign and Play in the Discourse of the Human Sciences', in R. Macksey and E. Donato (eds), *The Languages of Criticism and the Sciences of Man* (Baltimore: Johns Hopkins University Press, 1970), p 250.

81 *Ibid.*, p 267.

[82] *Ibid.*, p 266.

[83] *Ibid.*, p 267.

[84] R. Peierls, *Surprises in Theoretical Physics* (Princeton: Princeton University Press, 1979), p 24. Peierls immediately went on: 'Very probably some of the early opposition [to the theory of relativity] was not really based on the content of the new theory, which was difficult to grasp, but on the name.' Much of the postmodern use of relativity falls into the same error.

[85] G. Barker, *The True Confessions of George Barker* (London: MacGibbon & Kee, 1965), p 36.

[86] Published under the title, *Christian Morality* (Oxford: Clarendon Press, 1936).

[87] *Ibid.*, p 6.

[88] Henson, *Bishoprick Papers* (London: Oxford University Press, 1946), p 335.

Chapter 3. The Appeal to Value: Art in China and the West

[1] 'Black Iris II', in J. Tesch and E. Hollmann (eds), *Icons of Art: The Twentieth Century* (Munich: Prestel, n.d.), p 76.

[2] D. Hume, *An Enquiry Concerning the Principles of Morals*, 1751, App 1, 'Concerning Moral Sentiment', §3, in R.P. Wolff (ed), *The Essential David Hume* (New York: New American Library, 1969), pp 242f.

[3] B.R. Tilghman, *But Is It Art?* (Oxford: Blackwell, 1984).

[4] *Ibid.*, p 187.

[5] *Ibid.*, p 35.

[6] This is the title of Chapter 3 of S.R. Letwin's *The Pursuit of Certainty: David Hume, Jeremy Bentham, John Stuart Mill, Beatrice Webb* (Cambridge: Cambridge University Press, 1965), pp 29–40. This assumed conflict and its consequences are discussed further on pp 107–9.

[7] Hume, *Treatise of Human Nature*, ed L.A. Selby-Bigge (Oxford: Oxford University Press, 1955), p 469.

[8] Hume, *Enquiry*, §1, in R.P. Wolff (ed), *The Essential David Hume*, p 158.

[9] N. Nicholson, *The Lakers: The Adventures of the First Tourists* (London: Robert Hale, 1955), p 194.

[10] H. Osborne, *Theory of Beauty: An Introduction to Aesthetics* (London: Routledge & Kegan Paul, 1952), p 133.

[11] *Maya* is often translated as 'illusion', but in fact *maya* is the power of Brahman or of God to bring everything into the form in which it appears. Because everything is created and contingent, nothing is permanent. So *maya* takes on the nuance of illusion only because humans superimpose on appearances ignorant ideas of their own and take their mistakes to be real, as though one can find non-contingent realities in the midst of impermanent contingency. Of this error, the classic example is the rope on the path onto which we superimpose our false belief that it is a snake.

Notes to pages 43–46

As Gaudapada (eighth century CE) put it in his commentary on *Mundaka Upanishad*: 'A rope not clearly seen in the dark is imagined to be things like a snake or a trickle of water. The Self is misperceived in the same way. But when the rope is seen for what it is, false perceptions are dissolved, and consciousness becomes aware of non-duality with the recognition, This is only a rope. So it is with the discernment of what the Self is' (*Mandukyakarika*, 2.17f.). This is so basic to our constitution as human beings (to our being the sort of creatures that we are) that it is fundamental to the Indian understanding of what there is about us that needs to be dealt with if we are to escape from the shackles of time and of rebirth.

12   L. Weiskrantz, 'Behavioural Changes Associated with Ablation of the Amygdaloid Complex in Monkeys', *Journal of Comparative and Physiological Psychology*, XLIX (1956), pp 381–91.

13   For a summary, see E.T. Rolls, *The Brain and Emotion* (Oxford: Oxford University Press, 1999), pp 98–105.

14   See the review in J.P. Aggleton, *The Amygdala: Neurobiological Aspects of Emotion, Memory, and Mental Dysfunction* (New York: Wiley, 1992).

15   R. Joseph, *The Naked Neuron: Evolution and the Languages of the Brain and Body* (New York: Plenum, 1993), p 83.

16   C. Darwin, *The Expression of Emotion in Man and Animals* (London: Friedmann, 1979), p 38.

17   J.E. LeDoux, *The Emotional Brain* (London: Weidenfeld, 1998), p 174. Unpacked, LeDoux's argument is this: 'By way of the amygdala and its input and output connections, the brain is programmed to detect dangers, both those that were routinely experienced by our ancestors and those learned about by each of us as individuals, and to produce protective responses that are most effective for our particular body type, and for the ancient environmental conditions under which the responses were selected. Prepackaged responses have been shaped by evolution and occur automatically, or as Darwin pointed out, involuntarily. They take place before the brain has had the chance to start thinking about what to do. Thinking takes time, but responding to danger often needs to occur quickly and without much mulling over the decision. Recall Darwin's encounter with the puff adder at the Zoological Gardens – the snake struck and Darwin recoiled back quick as a flash. If the snake had not been behind glass, Darwin's life would have been at the mercy of his involuntary responses – if they were quick enough, he would have survived; if they were too slow, he would have perished. He certainly had no time to decide whether or not to jump once the snake started to strike. And even though he had resolved not to jump, he could not stop himself.'

18   Rolls, *The Brain and Emotion*, p 104.

Notes to pages 46–50

[19] LeDoux, 'Emotions: A View through the Brain', in R.J. Russell (ed), *Neuroscience and the Person* (Notre Dame: Notre Dame University Press, 1999), pp 101–17.

[20] *Ibid.*, p 115.

[21] H. Nisijo, T. Ono and H. Nishino, 'Single Neuron Responses in Amygdala of Alert Monkey during Complex Sensory Stimulation with Affective Significance', *Journal of Neuroscience*, VIII (1988), pp 3,570–83.

[22] F.A.W. Wilson and E.T. Rolls, 'The Primate Amygdala and Reinforcement: A Disassociation between Rule-based and Associatively Mediated Memory Revealed in Amygdala Neuronal Activity' (in preparation). The work is summarized in Rolls, 'Neurophysiology and Functions of the Primate Amygdala, and the Neural Basis of Emotion', in J.P. Aggleton (ed), *The Amygdala: A Functional Analysis* (Oxford: Oxford University Press, 2000), pp 447–78.

[23] A.L. Fairhall, G.D. Lewen, W. Bialek and R.R. de R. van Steveninck, 'Efficiency and Ambiguity in an Adaptive Neural Code', *Nature*, CCCCXII, no. 6849 (2001), pp 787–92.

[24] D.H. Lawrence, 'Art and Morality', in E.D. McDonald (ed), *Phoenix: The Posthumous Papers of D.H. Lawrence* (New York: Viking, 1972).

[25] B. Bryson, *The Lost Continent: Travels in Small Town America* (London: Abacus, 1996), p 231.

[26] *Ibid.*, p 232.

[27] A.C. Danto, *The Transfiguration of the Commonplace* (Cambridge, Mass.: Harvard University Press, 1981). As Danto explains (p v), the title occurred originally in Muriel Spark's novel, *The Prime of Miss Jean Brodie*, as a work by Sister Helena of the Transfiguration. The title was happily passed on by Muriel Spark.

[28] Bryson, *The Lost Continent*, p 230.

[29] For a summary of somatic exploration and exegesis, see my article, 'Religion', in *The Oxford Dictionary of World Religions*, pp xviif.

[30] Wittgenstein, in his lectures on aesthetics, questioned this on the ground that 'in real life, when aesthetic judgements are made, aesthetic adjectives such as "beautiful", "fine", etc., play hardly any role at all', except perhaps as interjections (C. Barrett (ed), *Lectures and Conversations on Aesthetics, Psychology and Religious Belief* (Oxford: Blackwell, 1966), p 3). But the lectures on aesthetics were given in 1938, long before this recent work in the neurosciences. Thus Wittgenstein assumed, right at the beginning of the lectures, that the word 'beautiful' is used to pick out a certain property or quality, 'that of being beautiful' – and Hume had dealt with that! Consequently, Wittgenstein argued that in aesthetic judgement (for example, in a poetical or a musical criticism), 'the words you use are more akin to "right" and "correct" (as these words are used in ordinary speech) than to "beautiful" and "lovely"' (p 3).

Notes to pages 51–54

Now, however, we understand why, given the way that our brains and bodies work, the vocabularies of beauty are necessary, and why they are different from those of being right and being correct. Wittgenstein designed a house for his sister, and she recorded how, when the building was nearly finished, he became dissatisfied with one room, and had the ceiling pulled down and then rebuilt 3cm. higher ('My Brother Ludwig', in B. Leitner, *The Architecture of Ludwig Wittgenstein* (New York: New York University Press, 1976), p 21). That is very different from John Ruskin standing amid 'the stones of Venice' and using 'the perfection of her beauty' to write solemn warnings to the people of England.

31   J.M. Whistler, 'Mr Whistler's Ten O'Clock', in *The Gentle Art of Making Enemies*, quoted in R. Aldington, *The Religion of Beauty* (London: Heinemann, 1950), pp 224, 229.

32   *Enso* (Japanese, 'circle') is the symbol of the absolute, uncontained and un-conditioned reality of the Buddha-nature. The execution of the *enso* is both an act of meditation and also the realization of the Buddha-nature, because it carries the one who paints it into the realization of that nature. It is usually painted in a single, flowing stroke of the brush, though there are examples of two lines – e.g. Bankei (1622–93), on whom Stephen Addiss commented: 'Bankei, ever the individualist, used two strokes, each strongly and quickly ar-ticulated. The effect is to give an entirely new meaning to the form; the strokes enclose each other like an embrace yet still suggest both emptiness and completion' (*The Art of Zen: Paintings and Calligraphy by Japanese Monks, 1600–1925* (New York: Harry N. Adams, 1998), p 73). Bankei (popular today in the West) was in any case a highly original teacher of the *fusho* (unborn mind), whose status as an authentically enlightened teacher was disputed by his con-temporaries: see H. Dumoulin, *Zen Buddhism: A History. II. Japan* (New York: Macmillan, 1990), pp 310–25. In either case, the recognition of value as a property in the circle is literally obvious. It may be obvious mainly to Zen Buddhists, but it still remains a property in the circle.

33   See G. Vasari, *Lives of the Painters, Sculptors and Architects*, trans G. du C. de Vere (London: Everyman's Library, 1996), p 103.

34   See G. Poulet, trans C. Dawson and E. Coleman, *The Metamorphoses of the Circle* (Baltimore: Johns Hopkins University Press, 1970).

35   Cicero, *De Natura Deorum*, 2.22.57.

36   In M. Kemp and M. Walker, *Leonardo on Painting: An Anthology of Writings by Leonardo da Vinci with a Selection of Documents Relating to his Career as an Artist* (London: Yale University Press, 1989), p 222.

37   C. Grayson, *Leon Battista Alberti on Painting and on Sculpture* (London: Phai-don, 1972), p 121.

38   C. Johnson, *The Language of Painting* (Cambridge: Cambridge University Press, 1949; London: Faber & Faber, 1971), p 23.

Notes to pages 54–59

[39] *Ibid.*, p 1.

[40] C. Rosen, *The Classical Style: Haydn, Mozart, Beethoven* (London: Faber & Faber, 1971).

[41] *Ibid.*, p 23.

[42] *Ibid.*, p 69.

[43] A famous, though perhaps notorious, example is Loran's analysis of the formal structure underlying the composition of Cézanne's paintings – notorious, because one of his diagrams was used by Roy Lichtenstein, 'Portrait of Madame Cézanne' (1963), and Loran charged him with plagiarism. See E. Loran, *Cézanne's Composition: Analysis of His Form with Diagrams and Photographs of his Motifs* (Berkeley: University of California Press, 1946).

[44] See especially J. Gage, *Colour and Meaning: Art, Science and Symbolism* (London: Thames and Hudson, 1999); *Colour and Culture: Practice and Meaning from Antiquity to Abstraction* (London: Thames and Hudson, 1995).

[45] Gage, *Colour and Culture*, p 268.

[46] H.B. Chipp, *Theories of Modern Art: A Source Book by Artists and Critics* (Berkeley: University of California Press, 1968); *Daily Telegraph*, 14 April 1989, p 12.

[47] E. Auerbach, *Mimesis* (Princeton: Princeton University Press, 1953).

[48] According to Frederic Stephens, a friend of Hunt, the painting was originally intended as a satire 'on the reported defenceless state of our country against foreign invasion'. It was renamed when the threat of invasion receded. See L. Lambourne, *Victorian Painting* (London: Phaidon, 1999), p 246.

[49] The painting is now in the Tate Gallery. For Ruskin's comment, see R. Hewison, *Ruskin, Turner and the Pre-Raphaelites* (London: Tate Gallery Publishing, 2000), p 205.

[50] Vasari, *Lives of the Painters*, vol 1, p 117.

[51] Much earlier, he wrote: 'it is the artist who is truthful and it is photography which lies, for in reality time does not stop, and if the artist succeeds in producing the impression of a movement which takes several moments for accomplishment, his work is certainly less conventional than the scientific image, where time is suspended.' Rodin to Paul Gsell, quoted in A.E. Elsen, *In Rodin's Studio: A Photographic Record of Sculpture in the Making* (Oxford: Phaidon, 1980), p 11.

[52] For the controversy, see J. Cladel, *Rodin: sa vie glorieuse, sa vie inconnue* (Paris: Grasset, 1950).

[53] 'The Old Courtesan', 1880–83, was subsequently called, 'The Helmet-Maker's Wife'.

[54] Quoted in Elsen, *In Rodin's Studio*, p 171.

[55] G. Eliot, *Adam Bede* (Oxford: World Classics, 1922), pp 196f.

Notes to pages 59–62

56 W. Holman Hunt, *Pre-Raphaelitism and the Pre-Raphaelite Brotherhood*, II, ch. 16, in D.S.R. Welland, *The Pre-Raphaelites in Literature and Art* (London: Harrap, 1953), p 69.

57 For an account which includes reproductions of three of the portraits, see 'The Portrait of Henry VIII', *Trinity College, Cambridge, Annual Record*, 2003, pp 17–19.

58 R. Strong, *Portraits of Queen Elizabeth I* (Oxford: Clarendon Press, 1963), p 34; the two references are to John Davies, 'To Her Picture' in *Hymnes to Astraea*: 'So dull her counterfeit should be,/And she so full of glory.' See also M. Jenkins, *The State Portrait, its Origin and Evolution*, Monographs on Archaeology and Fine Arts, iii (1947).

59 Strong, *Portraits of Queen Elizabeth I*, p 34.

60 M. Elwin (ed), *The Autobiography and Journals of Benjamin Robert Haydon* (London: Macdonald, 1950), p 648.

61 P.B. Shelley, letter to Maria Gladstone, 1 July 1820.

62 Elwin (ed), *The Autobiography and Journals of Benjamin Robert Haydon*, p 159.

63 *Ibid.*, pp 162f.

64 *Ibid.*, p 163.

65 O. Wilde, *The Picture of Dorian Gray*, in *Complete Works of Oscar Wilde* (London: Collins, 1981), p 17. There is a self-contradiction in this famous saying, because it occurs in the preface to a work that is 'a moral preachment against soulless hedonism'. Christopher Clausen made the point: 'Many critics today would agree with Wilde's formulation, which has come to represent one of the more visceral attitudes of twentieth-century literary study. The fact that Wilde made his remarks in the preface to one of the few successful moral allegories of modern times, *The Picture of Dorian Gray*, is an irony that usually goes unnoticed' (*The Moral Imagination: Essays on Literature and Ethics* (Iowa: University of Iowa Press, 1986), p ix).

66 A poem should be equal to:
    Not true

    For all the history of grief
    An empty door and a maple leaf

    For love
    The leaning grasses and two lights

    Above the sea –
    A poem should not mean
    But be
    (Archibald MacLeish, 'Ars Poetica', in *Streets in the Moon* (Boston: Houghton Mifflin, 1926)).

67 I.A. Richards, *Poetry and Sciences* (New York: W.W. Norton, 1970), p 33.

68 K. Clark, 'Art and Society', *The Cornhill Magazine*, Autumn 1960, p 324.

Notes to pages 63–67

[69] M. Loehr, *The Great Painters of China* (London: Phaidon, 1980), p 14; Chiang Yee, *The Chinese Eye: An Interpretation of Chinese Painting* (London: Methuen, 1935), p 33.

[70] 'The Poet and his Friends on the Riverbank under the Full Moon', attributed to Qiao Zhong-chang (twelfth century), in the John Crawford Collection, New York.

[71] M. Sullivan, *Symbols of Eternity: The Art of Landscape Painting in China* (Oxford: Clarendon Press, 1979), p 8.

[72] This led to the nineteenth-century condemnation of Chinese art in the West, epitomized in the *National Cyclopedia of Useful Knowledge* (London: Charles Knight, 1848), vol 4, p 497: 'With regard to the fine arts, the Chinese have not made much progress. In painting, their colours are beautiful, but their perspective is erroneous.'

[73] O. Sirén, *The Chinese on the Art of Painting* (Peking: Henri Vetch, 1936).

[74] H. Honour and J. Fleming, *A World History of Art* (London: Laurence King, 1999), p 856.

[75] Sullivan, *Symbols of Eternity*, p 80.

[76] O. Sirén, *Chinese Painting: Leading Masters and Principles* (London: 1958), vol 4, p 19.

[77] S.E. Lee and Wai-kam Ho, *Chinese Art under the Mongols: The Yüan Dynasty (1279–1368)* (Cleveland: Cleveland Museum of Art, 1968), no. 251; Sullivan, *Symbols of Eternity*, p 103.

[78] Wang Wei was an influential artist and teacher, whose *Preface on Painting* has been preserved only in a short and possibly altered form, in Zhang Yan-yuan's *Li-dai ming-hua-ji*, ch. 6. On the difference between imitation and art, he wrote: 'When speaking of painting, what we ultimately are looking for is nothing but the [expressive] power contained in it. Even the ancients did not make their pictures as mere topographic records of cities, districts and townships. ... Based on the shapes is an indwelling spiritual force [ling]; what activates that force is the mind [hsin]. If the mind ceases, that spiritual force is gone because what it relies on is not activated. The eye has its limitations; what it does see is not the whole (but only the exterior).' See M. Loehr, *The Great Painters of China*, pp 22f.

[79] S. Bush, *The Chinese Literati on Painting: Su Shih (1037–1101) to Tung Ch'I-ch'ang (1555–1636)* (Cambridge, Mass.: Harvard Yenching Studies, 1971), p 22.

[80] *Śunyata* is often translated by words like 'emptiness', 'the void', but it conveys something more like 'devoid of characteristics': appearances present themselves to perception as though they are substantial and real, but since they are impermanent and are manifestations of nothing but the Buddha-nature (p 67), they are really devoid of any characteristics that might differentiate them from each other or from anything else.

Notes to pages 67–72

To quote from my entry in *The Oxford Dictionary of World Religions*: 'In early Buddhism, the term is used primarily in connection with the "no-self" doctrine to denote that the Five Aggregates are "empty" of the permanent self or soul which is erroneously imputed to them. By extension the term came to be applied to reality as a whole: just as the individual is "void" of a self, in the sense of an unchanging controlling agency, so too is the whole universe "void of a self or anything belonging to a self". The doctrine of emptiness, however, received its fullest elaboration at the hands of Nagarjuna, who wielded it skillfully to destroy the substantialist conceptions of the Abhidharma schools of the Hinayana. The latter considered the emptiness of phenomena to lie in their impermanency, and maintained that while entities are subject to a process of almost instantaneous change, they are none the less substantial and possessed of a true "self-nature" in their moment of being. Nagarjuna took the doctrine of emptiness to its logical conclusion and argued that this notion of a "self-nature", albeit momentary, was at variance with the Buddha's teaching of no-self. The true nature of phenomena, he concluded, was to be empty of a self or self-essence of any kind. The doctrine of emptiness is the central tenet of the Madhyamaka school, and a statement of Nagarjuna's views in support of it may be found in his *Mulamadhyamakakarika*. Emptiness thus becomes a fundamental characteristic of Mahayana Buddhism. The teaching is subtle and its precise formulation a matter of sophisticated debate, since the slightest misunderstanding is said to obstruct progress towards final liberation. Emptiness is never a generalized vacuity, like an empty room, but always relates to a specific entity whose emptiness is being asserted. In this way up to twenty kinds of emptiness are recognized, including the emptiness of emptiness.'

[81] Sullivan, *Symbols of Eternity*, p 48.

[82] On this, see J.W. Bowker, *Hallowed Ground: Religions and the Poetry of Place* (London: SPCK, 1993), pp 7–13.

[83] B. Watson, *Chinese Rhyme-Prose: Poems in the Fu Form from the Han and Six Dynasties Periods* (New York: Columbia University Press, 1971), p 81.

[84] A.E. Housman, *The Name and Nature of Poetry* (Cambridge: Cambridge University Press, 1962).

[85] M. Tippett, *Music of the Angels: Essays and Sketchbooks* (London: Eulenburg Books, 1980), p 52.

[86] Hume, p 375.

[87] S. Hampshire, *Innocence and Experience* (London: Allen Lane, 1989), pp 64f.

Chapter 4. The Appeal to Value: Ethics and Human Behaviour

[1] Count von Eckstädt, Diary, quoted in J.G. Legge, *Rhyme and Revolution in Germany* (London: Constable, 1918), pp 267f.

[2] P.W. Wilson (ed), *The Greville Diary* (London: Heinemann, 1927), vol 2, p 586.

[3] Braley (ed), *Letters of Herbert Hensley Henson*, p 42.

Notes to pages 73–80

4 See p 39.

5 J. Lear, 'On Moral Objectivity', in S.C. Brown (ed), *Objectivity and Cultural Divergence* (Cambridge: Cambridge University Press, 1984), p 135.

6 R.M. Hare, 'Ontology in Ethics', in T. Honderich (ed), *Morality and Objectivity: A Tribute to J.L. Mackie* (London: Routledge & Kegan Paul, 1985), p 47.

7 B. Russell, *The Autobiography of Bertrand Russell* (London: Unwin Books, 1975), p 9.

8 B. Russell, 'Reply to Criticisms', in P.A. Schlipp (ed), *The Philosophy of Bertrand Russell* (La Salle: Open Court Publishing Company (Library of Living Philosophers), 1944), pp 720–5, quoted in C. Pigden (ed), *Russell on Ethics* (London: Routledge, 1999), p 148.

9 D. Cupitt, *The New Christian Ethics* (London: SCM, 1988).

10 *Ibid.*, p 67.

11 *Ibid.*, p 32.

12 *Ibid.*, p 34.

13 *Ibid.*, p 133.

14 *Ibid.*, p 27.

15 *Ibid.*, p 35.

16 *Ibid.*, p 3.

17 *Ibid.*, pp 10f., 33f.

18 *Ibid.*, p 84.

19 Cupitt has also objected to 'critical realism' on the grounds that the phrase is an oxymoron: 'I maintain that "critical realism" is an oxymoron, because consistently critical thinking shows that we are always inside our own vocabularies and our own angle on the world. We should give up the idea that we can somehow jump right out of our own limitations and achieve absolute knowledge, while yet remaining ourselves' (letter to *The Independent*, 11 August 1994). The fallacy described in Note 20 below (*either* being consistently critical *or* achieving absolute knowledge) vitiates his argument, since critical realists are not committed in any way to achieving absolute knowledge.

20 The fallacy, which appears so often in exam papers with the added instruction, 'Discuss', is described in my *Licensed Insanities* (p 101): 'It is the fallacy which poses a question as though only one of two alternatives (to the exclusion of any other additional possibility) is correct and has to be chosen. The fallacy is to ask, for example, "Concorde: white elephant or technological miracle?" The point is that it may be both those things – or neither – or indeed one but not the other – or much more besides.'

21 See, for example, *Evangelium Vitae* (London: Catholic Truth Society, n.d.), pp 126f.

22 Circulated copy of the lecture, p 3.

23 *Gaudium et Spes* (Second Vatican Council Pastoral Constitution on the Church in the Modern World), p 16.

Notes to pages 80–88

24 See, for example, A.P. d'Entrèves, *Natural Law* (London: Hutchinson, 1967), pp 8-10; D.J. O'Connor, *Aquinas and Natural Law* (London:, Macmillan, 1967), p 57: 'The concept of natural law is basic to the moral philosophy of Aquinas. But it has a very long history. The doctrine has taken many forms in the writings of philosophers and jurists from the Greeks to the present day. Indeed, these various versions of the doctrine differ so much both in their detail and in their philosophical bases that it is very misleading to talk of *the* theory of natural law. In so far as any common core can be found to the principal versions of the natural law theory, it seems to amount to the statement that the basic principles of morals and legislation are, *in some sense or other*, objective, accessible to reason and based on human nature. But so vague a specification tells us little until we decide the sense to be put upon these phrases. Many forms of the theory have, in addition, a theological foundation, and this is, in the case of St. Thomas, very important. But here too we have to ask how exactly the natural law is conceived to be related to God.'

25 Thomas Aquinas, *Summa Theologica*, 1a 2ae q.91 a.2.

26 *Veritatis Splendor* (London: Catholic Truth Society, 1993), §41.

27 D. Westley, *Morality and Its Beyond* (Mystic: Twenty-Third Publications, 1984), p 199.

28 John Paul II, *Crossing the Threshold of Hope*, p 52.

29 K. Grahame, *The Golden Age* (London: John Lane, 1915), p 241.

30 C. Moorehead, *Martha Gellhorn: A Life* (London: Chatto & Windus, 2003), p 284.

31 Rolls, *The Brain and Emotion*, p 67. In his criticism of the James–Lange theory of emotions, he made the same point: 'The theory leaves unanswered perhaps the most important issue in any theory of emotion: Why do some events make us run away (and then feel emotional), whereas others do not?' (p 71). See also J.W. Bowker, *The Sense of God*, where the consequent importance of cognition is stressed.

32 A.J. Calder et al, 'Facial Emotion Recognition after Bilateral Amygdala Damage: Differentially Severe Impairment of Fear', *Cognitive Neuropsychology*, XIII (1996), pp 699–745.

33 P. Ekman and W.V. Friesen, *Pictures of Facial Affect* (Palo Alto: Consulting Psychologists Press, 1976).

34 S.K. Scott et al, 'Impaired Auditory Recognition of Fear and Anger Following Bilateral Amygdala Lesions', *Nature*, CCCLXXXV (1997), pp 254–7.

35 *Independent*, 16 December 1992, p 17.

36 *Daily Telegraph*, 14 January 2004, p 7. For the original article, see V. Curtis, R. Aunger and T. Rable, 'Evidence that Disgust Evolved to Protect from Risk of Disease', *The Royal Society Biology Letters*, 2004.

Notes to pages 88–96

37 G.E.M. Anscombe, *The Collected Philosophical Papers of G.E.M. Anscombe: III. Ethics, Religion and Politics* (Minneapolis: University of Minnesota Press, 1981), p 139.

38 P. Foot, *Natural Goodness* (Oxford: Clarendon Press, 2001), p 24.

39 In M. Schofield et al (eds), *Doubt and Dogmatism: Studies in Hellenistic Epistemology* (Oxford: Clarendon Press, 1980), p 13.

40 *Independent*, 22 October, p 3.

41 H. Carpenter (ed), *The Letters of J.R.R. Tolkien* (Boston: Houghton Mifflin, 1981), p 243.

42 J.R.R. Tolkien, *The Lord of the Rings* (London: Folio Society, 1977), vol 2, p 213.

43 Carpenter (ed), *The Letters of J.R.R. Tolkien*, p 252.

44 Keehok Lee, *A New Basis for Moral Philosophy* (London: Routledge, 1985), pp 127f.

45 J.L. Mackie, *Ethics: Inventing Right and Wrong* (London: Penguin, 1977).

46 *Ibid.*, p 80.

47 C. Taylor, *Sources of the Self: The Making of Modern Identity* (Cambridge: Cambridge University Press, 1989), p 8.

48 Lee, *A New Basis for Moral Philosophy*, p 129.

49 *Ibid.*, pp 99f.,107.

50 'Polish bees headed straight back to their hives when they sensed contamination from the Soviet Union's Chernobyl disaster while the rest of Poland was still in the dark about the accident, according to a beekeeping expert. Mr Henry Ostach, who heads the Polish beekeepers' association, said apiarists were baffled when bees hid for several days after the explosion at the reactor. "When the explosion occurred, the bees interrupted their flight, although it was a fine sunny day. Not yet knowing anything about what had happened at Chernobyl, we wondered why the bees suddenly hid in their hives. They surrounded their queen very closely, beating their wings constantly in order to minimize the permeation of contamination"' (*The Times*, 14 February 1987, p 8).

51 T. Regan, *Defending Animal Rights* (Urbana: University of Illinois Press, 2001).

52 In E.F. Paul and J. Paul (eds), *Why Animal Experimentation Matters: The Use of Animals in Medical Research* (New Brunswick: Transaction Publishers, 2001).

53 For example, Michael Leahy rests his case 'against liberation' on 'an understanding of the true nature of conscious, appetitive beings, incapable of language' (Leahy, *Against Liberation: Putting Animals in Perspective* (London: Routledge, 1991), p 252).

54 Micah, 6.8.

55 Galatians, 5.22f.

Notes to pages 96–100

56 These are the first three of the conducive properties in Buddhism (as listed in the Abhidharma texts), leading to judgements of approval. As with Paul (in Galatians above), properties leading to an adverse judgement are also listed. In the Buddhist case, there are ten properties leading to judgements of approval, 18 leading to the opposite. A good property is known as *kusala-mahabhumika-dharma*. See D. Keown, *The Nature of Buddhist Ethics* (London: Macmillan, 1992), pp 62–3.

57 These are the seven properties in Indian Viśiṣṭadvaita which point to the ideal person. See T.W. Organ, *The Hindu Quest for the Perfection of Man* (Athens: Ohio University Press, 1970), p 169.

58 *Ren* is an impossible word to translate. For an illuminating introduction to its meaning, see Xinzhong Yao, *Confucianism and Christianity: A Comparative Study of Jen and Agape* (Brighton: Sussex Academic Press, 1996).

59 *Analects*, 17.6.

60 J.A. van der Ven, *Formation of the Moral Self* (Wheaton, Illinois: Eeerdmans, 1998). His book is an excellent example of the necessity to specify multiple constraints (pp 5–6) in the family, society, culture, education, tradition, economic context, etc., if we are to understand how or even whether the good person emerges in life.

61 E. Wyschogrod, *Saints and Post-modernism: Revisioning Moral Philosophy* (Chicago: University of Chicago Press, 1990).

62 J.W. Bowker, *Is God a Virus?*, p 238. The issue is reviewed in more detail in the section, 'Saints and the Denial of Humanity', pp 236–42.

63 G. Orwell, *The Lion and the Unicorn* (London: Penguin, 1982), p 32.

64 *The Listener*, 11 March 1943.

65 In *The Coming Crisis of Western Sociology*, Alvin Gouldner distinguished between background assumptions and domain assumptions, regarding the latter as 'background assumptions of more limited application': 'Domain assumptions are the background assumptions applied only to members of a single domain; they are in effect the metaphysics of a domain. Domain assumptions about man and society might include, for example, dispositions to believe that men are rational or irrational; that society is precarious or fundamentally stable; that social problems will correct themselves without planned intervention; that human behavior is unpredictable; that man's true humanity resides in his feelings and sentiments. I say that these "might" be examples of domain assumptions made about man and society, because whether they are or not is a matter that can be decided finally only by determining what people, including sociologists, believe about a given domain' (p 31).

66 I. Kant, *Foundations of the Metaphysics of Morals*, trans L.W. Beck (Indianapolis: Bobbs-Merrill, 1969).

Notes to pages 100–108

67   K. Jowitt, *New World Disorder: The Leninist Extinction* (Berkeley: University of California Press, 1992), p 307. The difference between Joshua and Genesis discourse is applied to John Paul II on p 281.

68   *The Times*, 30 July 2004, p 3. *The Times* added the note, 'Psychologists who have examined the effects of computer games on adolescents have found no definitive link with violent behaviour and called for more research' (2 August 2004, p 6).

69   *The Times*, 5 June 2002, p 8. He continued: 'It is the quality of commitment, commitment to the service of others. We know that you are, without falter or hesitation, totally committed to serving us, the British people. It is what drives you, you feel it, you let it guide your actions and it shows.'

70   See R. Dahl, *Boy: Tales of Childhood* (London: Penguin, 1992), pp 144–6.

71   D. McNaughton, *Moral Vision: An Introduction to Ethics* (Oxford: Blackwell, 1988).

72   Introduction to the retrospective exhibition, Tate Gallery, 1955, in J. Rothenstein (ed), *Stanley Spencer: The Man: Correspondence and Reminiscences* (London: Paul Elek, 1979), pp 149, 151.

73   N. Spivey, *Enduring Creation: Art, Pain and Fortitude* (London: Thames & Hudson, 2001), p 6.

74   M. Berenbaum, *The Vision of the Void: Theological Reflections on the Works of Elie Wiesel* (Middletown: Wesleyan University Press, 1979).

75   On the importance of downward causation in religious life, see J.W. Bowker, 'God, Spiritual Formation, and Downward Causation', *Theology*, CVII (2004), pp 81–8.

76   P. Byrne, *The Moral Interpretation of Religion* (Edinburgh: Edinburgh University Press, 1998), p 91. Byrne is discussing the apparent conflict between the world of values in a Kantian system and the world revealed by science and history: 'Religion embodies the faith that the two worlds can ultimately be reconciled' (p 91).

77   E. Kohak, *The Embers and the Stars: A Philosophical Inquiry into the Moral Sense of Nature* (Chicago: University of Chicago Press, 1984), p 85.

78   G.F. Woods, *A Defence of Theological Ethics* (Cambridge: Cambridge University Press, 1966), p 131.

79   P.G. Wodehouse, *The Girl in the Boat* (London: H. Jenkins, 1956).

80   A.N. Meltzoff and M.K. Moore, 'Imitation of Facial and Manual Gestures by Human Neonates' *Science*, CXCVIII (1977), pp 75–8.

81   J. Milton, *Paradise Lost*, 12.83, in E.H. Visiak (ed), *Milton: Complete Poetry and Selected Prose* (London: Nonesuch Library, 1952).

82   A.R. Damasio, *Descartes' Error: Emotion, Reason and the Human Brain* (New York: Quill, 2000), p 128.

83   J. Dryden, *Dedication of the Aeneis*.

Notes to pages 108–111

84 Milton, *Reason of Church-Government*, in Visiak (ed), *Milton: Complete Poetry and Selected Prose*, pp 559f.

85 Horace, *Epistolae*, 1.2.62.

86 *Ibid.*, 1.2.59–70: 'He who will not control his anger will wish undone what passionate resentment has prompted, when he hastens to gratify his feelings of insatiate hate. Anger is a brief fit of madness; rule your spirit which, unless it is under control, governs you: restrain it with bit and with chains. A master forms the docile colt with gentle neck to go on its way as the rider would direct it. The hunting dog, from the time that it barks at the deerskin hung in the hall, will hunt avidly in the woods.'

87 Anon., 'The Primate as Footballer', *The Boy's Own Annual*, XX (1898), p 415.

88 E. Waugh, *Decline and Fall*, ed D. Bradshaw (London: Penguin, 2000), p 64.

89 R. de Sousa, *The Rationality of Emotion* (Cambridge, Mass.: The MIT Press, 1987). Part of his argument is that emotions substitute for induction in ways that have been tested for general appropriateness through the process of evolution and natural selection, since strict induction would take too long (the snake would strike before we had worked out the best response); emotions thus precipitate us into action or judgement much faster, but not without *some* induction, not least by way of analogy.

90 Thomas Aquinas, *In Decem Libros Ethicorum Expositio*, 2082.

91 Hume, *Treatise*, 2.3.3.4. It is true that this statement was tempered in *An Enquiry Concerning the Principles of Morals* (so that 'reason and sentiment concur in almost all moral determinations and conclusions'), but the dynamic of the relationship remains.

92 B. Pascal, *Pensées*, 423, 424. Hume adopted the distinction between 'head and heart' in *An Enquiry Concerning the Principles of Morals*, in Wolff (ed), *The Essential David Hume*, p 258.

93 M. Scheler, *Schriften aus dem Nachlass*, I, *Zur Ethik und Erkenntnislehre* (Bern: Francke Verlag, 1957), p 362.

94 J. Maritain, *The Range of Reason* (New York: Scribner's, 1952), p 26.

95 J.J. Ratey, *A User's Guide to the Brain: Attention, Perception and the Four Theatres of the Brain* (New York: Little Brown, 2001), p 223.

96 Damasio, *Descartes' Error*, p 159.

97 Damasio, *The Feeling of What Happens: Body, Emotion and the Making of Consciousness* (London: Vintage, 2000), pp 51f.

98 *Ibid.*, p 159.

99 On this hypothesis, see Damasio, 'The Somatic Marker Hypothesis and the Possible Functions of the Prefrontal Cortex', *Philosophical Transactions of the Royal Society of London*, Series B (Biological Sciences), 351 (1996), pp 1413–20.

100 Rolls, *The Brain and Emotion*, pp 72f.

101 M.R. Bennett and P.M.S. Hacker, *Philosophical Foundations of Neuroscience* (Oxford: Blackwell, 2003), pp 212–16.

Notes to pages 112–113

[102] Rolls, *The Brain and Emotion*, pp 65f.

[103] Damasio, *The Feeling of What Happens*, p 41.

[104] See, for example, the discussion of dance (the 'dance' of bees, and classical ballet) in my *Is God a Virus?*, pp 64–7.

[105] W.J. Smith, in E.G. d'Aquili (ed), *The Spectrum of Ritual: A Biogenetic Structural Analysis* (New York: Colombia University Press, 1979), p 76. The point is also made strongly by Bennett and Hacker: 'We share many emotions with other animals – for example, curiosity, fear and anger. But the scope of possible objects of human curiosity, fear or anger vastly outstrips that of mere animal curiosity, fear and anger. The cognitive and appraisive aspects of human emotions reach far wider than those of animal emotions. Of course, the monkey that screams in fear of a snake knows snakes to be harmful. The lion that snarls in anger at the cub that is pestering it apprehends the cub as annoying. The pet cat that purrs with delight as its food is being prepared knows that it is about to be given its meal. But the cognitive capacities of animals are strictly limited by their lack of language. Many of the kinds of beliefs that enter into an account of human emotion are not beliefs that could possibly be ascribed to an animal. Emotions, we have suggested, are ways in which we manifest what is important to us. But human beings characteristically reflect on what matters to them, whereas non-language-using animals merely manifest what they care about (their territory, possession of the prey they have killed, their dominance in their group, etc.) in their behaviour. Hence the motivating power of an emotion in a non-language-using animal is both very restricted and importantly different from the motivating force of emotions among human beings. For human beings act for reasons, whereas animals at the most do so in a limited and attenuated sense. Human emotions colour thoughts and fantasies, but non-language-using animals do not dwell in thought upon the objects of their emotions and do not fantasize about the fulfillment of their hopes and fears – they lack the conceptual equipment which makes such thoughts and fantasies possible. Finally, there are many emotions, such as feelings of guilt, awe, remorse and moral indignation, which it is not logically possible for non-language-using animals to have. For such emotions presuppose mastery of a language and possession of appropriate concepts. An awareness of the limitations on animal emotion and its objects should provide a brake on unwarranted generalization in which conclusions derived from experiments on animals are extended without more ado to human beings. So neuroscientific research that focuses upon conditioned fear in rats screens out most of what is distinctive of human fears, in particular, and emotions in general' (Bennett and Hacker, *Philosophical Foundations of Neuroscience*, p 206).

Notes to pages 114–120

[106] Redundant in the good sense of the word. Freud argued that religious rituals, in their repetitiveness, are a form of pathologically obsessive behaviour made socially acceptable. However, redundancy is essential in information theory, and in the communication of information: without redundancy, no communication is possible. Thus rituals are highly redundant ways of protecting and transmitting information, above all how people have learnt through successive generations, to negotiate the transitions of time. See V.P. Gay, 'Reductionism and Redundancy', *Zygon*, XIII (1978), p 175.

[107] K.E. and J.M. Paige, *The Politics of Reproductive Ritual* (Berkeley: University of California Press, 1981).

[108] M. Bloch, *Prey into Hunter: The Politics of Religious Experience* (Cambridge: Cambridge University Press, 1992).

[109] A. Glucklich, *Sacred Pain: Hurting the Body for the Sake of the Soul* (Oxford: Oxford University Press, 2001), p 152.

[110] I. Weyrauther, *Muttertag und Mutterkreuz: Der Kult um die 'deutsche Mutter' im Nationalsozialismus* (Frankfurt am Main, 1993), p 162.

[111] Quoted in C. Silvester, *The Literary Companion to Parliament* (London: Sinclair-Stevenson, 1996), p 120.

Chapter 5. The Appeal to Coherence

[1] W.J. Friedman, *About Time: Inventing the Fourth Dimension* (Cambridge, Mass.: MIT Press, 1990), p 10.

[2] J. Boucher, '"Lost in a Sea of Time": Time-parsing and Autism', in C. Hoerl and T. McCormack (eds), *Time and Memory: Issues in Philosophy and Psychology* (Oxford: Clarendon Press, 2001), pp 111–35.

[3] *The Oxford Dictionary of World Religions*. See also B.J. Verkamp, *The Indifferent Mean: Adiaforism in the English Reformation to 1554* (Athens: Ohio University Press, 1977).

[4] 'Where religions are involved in conflict ..., there are some elementary guidelines to bear in mind, which are all the more important if the United Nations is to continue to be involved in peace-keeping or peace-making exercises.

'The first is to establish who is actually involved, and who, in the background, is offering support. There are no such people as "Muslims", "Christians", "Hindus", and so on – religious life is always more detailed, specific and complicated than that, as the reference to subsystems has tried to indicate. However, on the other hand, if the conflict is inter-religious, those involved will often identify themselves with the entire system, and think of themselves as Muslims, Christians, Hindus, and so on. The actual relation between subsystem and system may well be decisive in suggesting appropriate action.

Notes to page 120

'So, the second is to ask (and find out) how the issue in question looks from within the subsystems of the religions involved; and to bear in mind how it looks from the viewpoint of the overarching system, which may offer, at least theoretically, some controls. Thus, the IRA, despite the context and affiliations of its members, has been increasingly distanced from the Roman Catholic Church. For the outsider, there is no Archimedean ground on which to stand in order to issue superior judgements. No doubt, as outsiders, we will make our own judgements, but they will not fly nearer the sun than the birds already in the air.

'The third, therefore, is to learn to argue within the logic of the systems involved. This requires a great deal more than a knowledge of the relevant languages, although, obviously, that is a necessary condition of any conversation. It should then be possible to see what the logic of each system demands – or forbids – in relation to a particular issue. Logic rarely wins an argument; but it establishes what ought to be relevant to it.

'The fourth is to define, from the respective points of view of those involved, what the point or purpose or cause of this particular issue is, remembering then to set it in a wider network of constraints, since it is often there that the true root of intransigence is to be found.

'The fifth is to bear in mind that many people would rather die than abandon their faith ..., and that they are, consequently, prepared to die, and carry others with them into death, in ways that seem incredible to an outsider. Differences are profound and cannot be negotiated away as semantic inadequacies. It is essential to understand how a religion arrives at non-negotiable truth (i.e., at what counts as non-negotiable truth from its own point of view) according to the limits set within this tradition on legitimate or appropriate utterance. Otherwise, all that will be heard will be either the vague expressions of goodwill which bring interreligious meetings into such disrepute; or nothing.

'The sixth is to establish, in any conflict in which religions are involved, what the perceived conditions of continuity are for the participants involved: on what terms do they perceive that there is a reasonable chance that their grandchildren might grow up as Shiʻa Muslims, Vaishnava Hindus, Orthodox Jews, or whatever (bearing in mind the first point above about the relation between subsystems and systems)? No one knows the future. But this is not an exercise in exact prediction. Rather, it is a way of eliciting participant perception, by asking the following question in a practical way: "What do you think you need in order to secure the continuity of your lifeway, not just for tomorrow, but into some reasonable future?" Usually when this question is asked, the answer comes back, "Everything". But, at the other end of the spectrum, the Jews in the camps of Europe required virtually nothing to retain their Jewish identity and their faith.

Notes to page 120

Somewhere between the two extremes, between everything and nothing, there will be a compromise which is worth the risk; and if the word "compromise" is an unhappy one – a good umbrella but a poor roof, as Lowell called it – then take instead Samuel Johnson's "reciprocal concessions".

'The seventh is to apply the argument in Part I of this book [i.e. Part I of *Is God a Virus?*; see also pp 5–6 in this book]: talk not of causes, talk, rather, of constraints. This removes from the arena contest about "who caused" the conflict: in a network of constraints, responsibility is diffused and face is saved.

'The eighth is to remember, in the religious case, always, at some point (preferably early on), to follow Thomas Hardy's advice and take a full look at the worst. Religions can be extremely bad news, especially for those of whom they disapprove. However, at the same time, they are the source of immense goodness; faith is constantly moving mountains a little to the right, making the possible happen and the impossible take a little less long. The aim has to be to get the goodness of a system to engage with the evils the system is endorsing, or allowing, or ignoring, or setting forward – goodness and evil, that is, on its own terms of reference. What rapidly becomes apparent is that the definitions are not actually relative, and that there is a very extensive religious consensus concerning both good and evil. Religions *can* be allies in healing the Earth. Perhaps what is needed is not so much an Amnesty International as an Uposatha International (Uposatha is the fortnightly meeting when Buddhist bhikkhus, monks, review their offences against the basic code of behaviour, the Patimokkha – the review is against their own code, since there is no neutral Archimedean point on which to stand). What religions might one day attempt is a mutual reinforcement of each other in attempting to secure the behaviours which are appropriate, or moral, or good according to their own system. For example, the Qur'an states firmly, "There shall be no compulsion in religion" [2.256/7]. *Shari'a* (Muslim law) makes specific provision for non-Muslims who are "peoples of the Book" to practise their faith. If this does not happen, the protest against the betrayal of God and of his messenger can only effectively come from within dar al-Islam, the house of Islam. However, if other religions reinforce this protest, it can only be if they accept a comparable critique of the betrayals in their own case – betrayals, not in a generalized sense, but in the terms of appropriate conduct which lie in their own sources of constraint.

'Religions are powerful resources in the enterprise of being good, generous, and just, despite all there is in their present behaviours and past histories which contradict that. Unless we find some collective means to take this "full look at the worst", without pulling down the house all around us in the first moments of outrage, we will never reach the point where the resources of goodness and reconciliation can be mobilized together.

Notes to pages 120–128

Religions will continue to exhibit exactly what Roger Woddis wrote about, of
Europe, in a poem called "Europhoria":

> Gloria, gloria, Europhoria!
> Common faith and common goal!
> Meat and milk and wine and butter
> Make a smashing casserole!
> Let the end of all our striving
> Be the peace that love promotes,
> With our hands in perfect friendship
> Firmly round each other's throats' (pp 188–91).

5   Diary, quoted in F.A. Iremonger, *William Temple: His Life and Letters* (London:
Oxford University Press, 1949), p 396.

6   *Ibid.*, p 395. It eventually appeared as *The Christian Message in Relation to Non-
Christian Systems: Report of the Jerusalem Meeting of the International Missionary
Council* (London: Oxford University Press, 1928), vol 1.

7   Iremonger, *William Temple*, p 399.

8   *Ibid.*, p 396.

9   W. Temple, *Readings in St. John's Gospel* (London: Macmillan, 1955), p 320.

10 *Ibid.*, p 237.

11 *Ibid.*, p 320.

12 The first of the Thirty-Nine Articles of the Church of England reads: 'There is
but one living and true God, everlasting, without body, parts, or passions; of
infinite power, wisdom, and goodness; the Maker, and Preserver of all things
both visible and invisible. And in unity of this Godhead there be three Persons, of one substance, power, and eternity; the Father, the Son, and the Holy
Ghost.'

13 This is one group of the four groups of *avyakata* questions, questions which
cannot be answered and which therefore it is not profitable to pursue. They
are: whether the world is eternal, or not, or both, or neither; whether the
world is finite, or not, or both, or neither; whether the Tathagata [the Buddha
who has attained enlightenment] exists after death, or not, or both, or neither;
whether the soul is identical with the body, or not. For a discussion in relation to the development of Madhyamaka, see T.R.V. Murti, *The Central
Philosophy of Buddhism: A Study of the Madhyamika System* (London: Allen &
Unwin, 1955), pp 36f.

14 R. Walker, *The Coherence Theory of Truth: Realism, Antirealism, Idealism* (London: Routledge, 1989), p 2.

15 *Ibid.*

16 *Ibid.*, p 210.

17 L. BonJour, *The Structure of Empirical Knowledge* (Cambridge, Mass.: Harvard
University Press, 1985), p 108.

18 *Ibid.*, p 92.

Notes to pages 128–139

[19] The moth *biston betularia* is a standard example of natural selection in the present supporting the inference of the same process in the past: before the Industrial Revolution in England, the moth was light coloured; the settlement of soot on trees made the light-coloured moths easy prey, so that during the second half of the nineteenth century, natural selection ensured that the dark-coloured moths survived and became the more common. After the passing of clean air acts, soot disappeared in cities, and the light-coloured moths became the more common once more.

[20] E. Crispin (ed), *Best SF: Science Fiction Stories* (London: Faber & Faber, 1954), p 9.

[21] As, for example, in a production of *The Hound of the Baskervilles* in 2002, in which Holmes and Watson rightly get onto a train in Paddington in order to travel to the West Country, only to be seen immediately afterwards travelling east with St Paul's Cathedral in the background. In the prolonged correspondence that followed in *The Times*, Mr E.W. Lighton was even more disturbed to see the passengers alighting (again correctly) at Exeter, but from 'carriages drawn by a shunting tank engine bearing the initials SR'. A further correspondent, Keith Evans, made the point that lapses in continuity are 'a long cinematic tradition': 'In the Hollywood version of *Terror by Night* ... Holmes and Dr Watson ... depart from London Euston (Euston?) for Edinburgh hauled by an appropriate LMS Royal Scot locomotive. This shortly afterwards becomes what appears to be a Great Western King as it picks up water from that railway's Goring troughs in Berkshire, then transmutes in turn into an American express locomotive, a scale model of a Great Western 4-6-0 which crosses the same viaduct at least three times, and finally arrives at its Scottish destination as a German Pacific presumably misrouted from Hamburg or Berlin. At least the guard's cap remains reassuringly lettered LMS throughout' (*The Times*, Letters, 30 December 2002; 1 and 2 January 2003).

[22] Carpenter (ed), *The Letters of J.R.R. Tolkien*, p 277.

[23] *Ibid.*, pp 283f.

[24] E. Mendelson (ed), *W.H. Auden: Collected Poems* (London: Faber & Faber, 1994), p 157.

[25] J. Mumford, *The Question of Questions, Which, Rightly Resolved, Resolves All Our Questions in Religions* (London: Richards, 1841).

[26] J.W. Bowker, *What Muslims Believe* (Oxford: Oneworld, 1998), p 157.

[27] On these, see A. Hasan, *The Early Development of Islamic Jurisprudence* (Islamabad: Islamic Research Institute, 1970), pp 155–77.

[28] C.F.H. Henry and W.S. Mooneyham (eds), *One Race, One Gospel, One Task* (Minneapolis: World Wide Publications, 1967), vol 2, pp 191f.

[29] J.W. Bowker, *God: A Brief History*, p 186.

Notes to pages 139–147

[30] The difference between context-independent commands and context-dependent applications is massively important in the kind of moral decision making that is based on any text such as Scripture, which is believed to be revealed. It rules out using texts as 'moral bricks', although that is in fact what is repeatedly done. The meaning of this distinction is discussed in my *Is God a Virus?*, pp 54f., 252–4.

[31] O.M.T. O'Donovan, *Resurrection and Moral Order* (Leicester: Inter-Varsity Press, 1986), p 200.

[32] J. Hamilton, *God, Guns and Israel* (Stroud: Sutton Publishing, 2004).

[33] J.W. Bowker, *A Year to Live* (London: SPCK, 1991), p 173.

[34] J. MacLaughlin, *Is One Religion as Good as Another?* (London: Burns & Oates, 1891), p 235.

[35] Ryusaku Tsunoda et al (eds), *Sources of Japanese Tradition* (New York: Columbia University Press, 1964), vol 2, p 140.

[36] J. Hick, *An Interpretation of Religion* (London: Macmillan, 1989), p 36.

[37] The three options are lucidly described by A. Race, *Christians and Religious Pluralism: Patterns in the Christian Theology of Religions* (London: SCM Press, 1983).

[38] J.W. Bowker, *Is God a Virus?*, pp 174–83. A comparable view was developed later by S.M. Heim, *Salvations: Truth and Difference in Religion* (Mayknoll: Orbis Books, 1995).

[39] And about 400 million different bacteria and archaes: see C. Tudge, *The Variety of Life: A Survey and Celebration of All the Creatures That Have Ever Lived* (Oxford: Oxford University Press, 2000).

[40] C.M. Wilson (Lord Moran), *Winston Churchill: The Struggle for Survival, 1940–1965* (London: Constable, 1966), p 508.

[41] R. Niebuhr, 'The Moral World of John Foster Dulles', *New Republic*, 1 December 1958, p 8.

[42] I. Stravinsky, *The Poetics of Music in the Form of Six Lessons* (Harvard: Harvard University Press, 1977), pp 61f. On this understanding of constraint as the condition of freedom, see J.W. Bowker, *A Year to Live*, pp 1–7, 54–5, 116; and 'Religions as Systems', in the Report of the Doctrine Commission of the Church of England, *Believing in the Church* (London: SPCK, 1981), pp 159–89.

[43] W.R. Ashby, *An Introduction to Cybernetics* (London: Methuen, 1964), p 130.

[44] A. Young, *Out of the World and Back* (London: Hart-Davis, 1961), p 71.

[45] 'The most subtle and compelling of all temptations is the temptation of the inner circle. Men will lie, betray their wives and friends for admission to that circle', quoted in J. Ezard, review of *C.S. Lewis at the Breakfast Table*, *Guardian*, 21 August 1980, p 9.

[46] G. Eliot, *The Mill on the Floss* (London: T. Nelson, The Nelson Classics, n.d.), pp 560f.

47 Prestige, *The Life of Charles Gore*, p 398.

48 *Ibid.*, p 398.

Chapter 6. Sex and Safety: A New Crisis Facing Religions

1   Further information about Gresham College and its programmes of lectures can be obtained from Gresham College, Barnard's Inn Hall, Holborn, London, EC1N 2HH; e-mail: enquiries@gresham.ac.uk. In accord with Gresham's intention to make the latest research and knowledge known freely, transcripts of all recent lectures can be obtained from the same address, either in hard copy, or together on a free CD-ROM disk.

2   H. Dumoulin, *Zen Buddhism: A History, II, Japan*, p 260.

3   *Akal Ustat*, in N.G. Kaur Singh, *The Name of My Beloved: Verses of the Sikh Gurus* (London: HarperCollins, 1995), p 8.

4   For a summary, see my *God: A Brief History*, pp 340–1.

5   M. Mujeeb, *The Indian Muslims* (New Delhi: Munshiram Manoharlal, 1995), pp 244f.

6   The meaning of this phrase is uncertain. It is often taken to mean, 'in token of willing submission'.

7   Quran, 9.29.

8   Jihad is governed by strict rules derived from the Quran and from the Sunna (i.e. from Muhammad and his Companions) which neither the Taleban nor alQaeda observe. I have translated the rules in 'Jihad and Shari'a' in my *World Religions* (London: Dorling Kindersley, 2nd edition, 2003), pp 184–5.

9   R. Hooker, *The Works ...* (Oxford: Clarendon Press, 1885), vol 1, p 417.

10 *Ibid.*, p 421.

11 A. Harvey, *Light upon Light: Inspirations from Rumi* (Berkeley: North Atlantic Books, 1996).

12 Isolating genes in this way, as in the phrase (and book title) 'the selfish gene' is very misleading. One might equally write of the selfish protein. Nevertheless, it makes the point in a simple way. For a more careful discussion, see my *Is God a Virus?*, Part I.

13 J. O'Donohue, 'The Scourging at the Pillar', in *Conamara Blues* (Sydney: Bantam Books, 2001), p 75.

14 A.E. Rasa, C. Vogel and E. Voland, *The Sociobiology of Sexual and Reproductive Strategies* (London: Chapman and Hall, 1989), p xi.

15 *Ibid.*, p xi.

16 See V. Sommer, 'Infanticide among Free-Ranging Langurs (*Presbytis entellus*) at Jodhpur (Rajasthan/India): Recent Observations and a Reconsideration of Hypotheses', *Primates*, XXVIII (1987), pp 163–97.

17 N. Barley, *A Plague of Caterpillars* (London: Penguin, 1987), pp 98f.

18 J.W. Bowker, *What Muslims Believe*.

19 *Ibid.*, p 132.

Notes to pages 161–169

[20]   Quran, 2.183.

[21]   Quran, 2.229.

[22]   D.R. Brooks, *Auspicious Wisdom: The Texts and Traditions of Śrividya Śakta Tantrism in South India* (New York: State University of New York Press, 1992), p 61.

[23]   J.W. Bowker, *Worlds of Faith: Religious Belief and Practice in Britain Today* (London: Ariel Books, 1983), p 213.

[24]   *Women's Own*, 31 October 1987.

[25]   *Humanae Vitae*, §11.

[26]   *Ibid.*, §14, *Catechism of the Catholic Church* §2370.

[27]   *Catechism of the Catholic Church* §1954. On natural law, see also pp 79–82.

[28]   A.R.I. Doi, *Shari'ah: The Islamic Law* (London: Ta Ha, 1984), p 11.

[29]   On the different ways in which religions respond to the naturalization of death in our time, see J.W. Bowker, 'Die menschliche Vorstellung vom Tod', in C. von Barloewen (ed), *Der Tod in den Weltkulturen und Weltreligionen* (Munich: Diederichs, 1996), pp 406–32.

BIBLIOGRAPHY

Abbott, W.J., *The Documents of Vatican II* (London: Geoffrey Chapman, 1967)

Addiss, S., *The Art of Zen: Paintings and Calligraphy by Japanese Monks, 1600–1925* (New York: Harry N. Adams, 1998)

Aggleton, J.P., *The Amygdala: Neurobiological Aspects of Emotion, Memory, and Mental Dysfunction* (New York: Wiley, 1992)

——, *The Amygdala: A Functional Analysis* (New York: Wiley, 2000)

Aldington, R., *The Religion of Beauty* (London: Heinemann, 1950)

Anon., 'The Primate as Footballer', *The Boy's Own Annual*, XX (1898)

Anscombe, G.E.M., *The Collected Philosophical Papers of G.E.M. Anscombe: III, Ethics, Religion and Politics* (Minneapolis: University of Minnesota Press, 1981)

Aquinas, Thomas, *In Decem Libros Ethicorum Expositio*

——, *Summa Theologica*

Ashby, W.R., *An Introduction to Cybernetics* (London: Methuen, 1964)

Auerbach, E., *Mimesis: The Representation of Reality in Western Literature*, trans W.R. Trask (Princeton: Princeton University Press, 1953)

Barker, G., *The True Confessions of George Barker* (London: MacGibbon & Kee, 1965)

Barley, N., *A Plague of Caterpillars* (London: Penguin, 1987)

Bennett, M.R. and P.M.S. Hacker, *Philosophical Foundations of Neuroscience* (Oxford: Blackwell, 2003)

Berenbaum, M., *The Vision of the Void: Theological Reflections on the Works of Elie Wiesel* (Middletown: Wesleyan University Press, 1979)

Bloch, M., *Prey into Hunter: The Politics of Religious Experience* (Cambridge: Cambridge University Press, 1992)

Boas, F., 'Rudolph Virchow's Anthropological Work', *Science*, XVI (1902)

——, *Race, Language and Culture* (Chicago: University of Chicago Press, 1982)

BonJour, L., *The Structure of Empirical Knowledge* (Cambridge, Mass.: Harvard University Press, 1985)

Boothby, R.J.G., *Recollections of a Rebel* (London: Hutchinson, 1978)

Boucher, J., '"Lost in a Sea of Time": Time-Parsing and Autism', in C. Hoerl and T. McCormack (eds), *Time and Memory: Issues in Philosophy and Psychology* (Oxford: Clarendon Press, 2001)

Bowker, J.W., 'Religions as Systems', in the Report of the Doctrine Commission of the Church of England, *Believing in the Church* (London: SPCK, 1981)

——, 'Religious Studies and the Languages of Religions', *Religious Studies*, XVII (1981)

——, 'Only Connect ...', *Christian*, VII (1982)

—, *Worlds of Faith: Religious Belief and Practice in Britain Today* (London: Ariel Books, 1983)

—, *Licensed Insanities: Religions and Belief in God in the Contemporary World* (London: DLT, 1987)

—, 'The Religious Understanding of Human Rights and Racism', in D.D. Honoré (ed), *Trevor Huddleston: Essays on His Life and Work* (Oxford: Oxford University Press, 1988)

—, *A Year to Live* (London: SPCK, 1991)

—, *Hallowed Ground: Religions and the Poetry of Place* (London: SPCK, 1993)

—, *Is God a Virus? Genes, Culture and Religion* (London: SPCK, 1995)

—, *The Sense of God: Sociological, Anthropological and Psychological Approaches to the Origin of the Sense of God* (Oxford: Oneworld, 2nd edition, 1995)

—, 'Die menschliche Vorstellung vom Tod', in C. von Barloewen (ed), *Der Tod in den Weltkulturen und Weltreligionen* (Munich: Diederichs, 1996)

— (ed), *The Oxford Dictionary of World Religions* (Oxford: Oxford University Press, 1997)

—, 'Science and Religion: Contest or Confirmation?', in F. Watts (ed), *Science Meets Faith* (London: SPCK, 1998)

—, *What Muslims Believe* (Oxford: Oneworld, 1998)

—, *God: A Brief History* (London: Dorling Kindersley, 2002)

—, 'God, Spiritual Formation, and Downward Causation', *Theology*, CVII (2004)

Bowker, M., 'Lincolnshire 1536: Heresy, Schism or Religious Discontent?', in D. Baker (ed), *Studies in Church History: Schism, Heresy and Religious Protest* (Cambridge: Cambridge University Press, 1972)

Braley, E.F. (ed), *Letters of Herbert Henry Henson* (London: SPCK, 1951)

Brooks, D.R., *Auspicious Wisdom: The Texts and Traditions of Śrīvidyā Śākta Tantrism in South India* (New York: State University of New York Press, 1992)

Bruce, V. and A. Young, *In the Eye of the Beholder: The Science of Face Perception* (Oxford: Oxford University Press, 1998)

Bryson, B., *The Lost Continent: Travels in Small Town America* (London: Abacus, 1996)

Bush, S., *The Chinese Literati on Painting: Su Shih (1037–1101) to Tung Ch'I-ch'ang (1555–1636)* (Cambridge, Mass.: Harvard University Press, 1971)

Byrne, P., *The Moral Interpretation of Religion* (Edinburgh: Edinburgh University Press, 1998)

Calder, A.J. et al, 'Facial Emotion Recognition after Bilateral Amygdala Damage: Differentially Severe Impairment of Fear', *Cognitive Neuropsychology*, XIII (1996)

Caputo, J.D. (ed), *The Religious* (Oxford: Blackwell, 2002)

Carlyle, T., Letter to Mrs David Ogilvy, 25 July 1851

Carpenter, H. (ed), *The Letters of J.R.R. Tolkien* (Boston: Houghton Mifflin, 1981)

Catechism of the Catholic Church (London: Geoffrey Chapman, 1994)

Chipp, H.B., *Theories of Modern Art: A Source Book by Artists and Critics* (Berkeley: University of California Press, 1968)

Cicero, *De Natura Deorum*

Cladel, J., *Rodin: sa vie glorieuse, sa vie inconnue* (Paris: Grasset, 1950)

Clark, K., 'Art and Society', *The Cornhill Magazine*, Autumn 1960

Clausen, C., *The Moral Imagination: Essays on Literature and Ethics* (Iowa: University of Iowa Press, 1986)

Collins, H. and T. Pinch, *The Golem: What You Should Know about Science* (Cambridge: Cambridge University Press, 1998)

Coveney, P. and R. Highfield, *Frontiers of Complexity: The Search for Order in a Chaotic World* (London: Faber & Faber, 1996)

Crispin, E. (ed), *Best SF: Science Fiction Stories* (London: Faber & Faber, 1954)

Croce, B., *History: Its Theory and Practice*, trans D. Ainslee (New York: Harcourt Brace, 1923)

Cupitt, D., *The New Christian Ethics* (London: SCM, 1988)

Curtis, V., R. Aunger and T. Rable, 'Evidence that Disgust Evolved to Protect from Risk of Disease', *The Royal Society Biology Letters*, 2004

Dahl, R., *Boy: Tales of Childhood* (London: Penguin, 1992)

Damasio, A.R., 'The Somatic Marker Hypothesis and the Possible Functions of the Prefrontal Cortex', *Philosophical Transactions of the Royal Society of London*, Series B (Biological Sciences), 1996

—, *Descartes' Error: Emotion, Reason and the Human Brain* (New York: Quill, 2000)

—, *The Feeling of What Happens: Body, Emotion and the Making of Consciousness* (London: Vintage, 2000)

Danto, A.C., *The Transfiguration of the Commonplace* (Cambridge, Mass.: Harvard University Press, 1981)

d'Aquili, E.G., *The Spectrum of Ritual: A Biogenetic Structural Analysis* (New York: Columbia University Press, 1979)

Darwin, C., *The Origin of Species* (New York: New American Library, 1958)

—, *The Expression of Emotion in Man and Animals* (London: Friedmann, 1979)

Davies, J., 'To Her Picture' in *Hymns to Astraea*

Dawkins, R., *The Selfish Gene* (Oxford: Oxford University Press, 1976, 1989)

—, Interview in the *Daily Telegraph*, 31 August 1992

de Sousa, R., *The Rationality of Emotion* (Cambridge, Mass.: MIT Press, 1987)

Demosthenes, *Kata Mediou*, in S.H. Butcher (ed), *Demosthenis Orationes* (Oxford: Clarendon Press, 1907)

d'Entrèves, A.P., *Natural Law* (London: Hutchinson, 1967)

Derrida, J., 'Structure, Sign and Play in the Discourse of the Human Sciences', in R. Macksey and E. Donato (eds), *The Languages of Criticism and the Sciences of Man* (Baltimore: Johns Hopkins Press, 1970)

Doctrine Commission of the Church of England, *Believing in the Church* (London: SPCK, 1981)

Dodds, M.H. and R., *The Pilgrimage of Grace 1536–1537 and the Exeter Conspiracy, 1538* (Cambridge: Cambridge University Press, 1915)

Doi, A.R.I., *Shari'ah: The Islamic Law* (London: Ta Ha, 1984)

Dryden, J., *Vergil: The Aeneid* (New York: Limited Editions Club, 1944)

Dumoulin, H., *Zen Buddhism: A History, II, Japan* (New York: Macmillan, 1990)

Earman, J. and C. Gilmour, C., 'Relativity and Eclipses: The British Eclipse Expeditions and Their Predecessors', in *Historical Studies in the Physical Sciences*, XI (1980)

Ekman, P. and W.V. Friesen, *Pictures of Facial Affect* (Palo Alto: Consulting Psychologists Press, 1976)

Eliot, G., *Adam Bede* (Oxford: World Classics, 1922)

—, *The Mill on the Floss* (London: T. Nelson, n.d.)

Eliot, T.S., *Four Quartets* (London: Faber & Faber, 1944)

Elsen, A.E., *In Rodin's Studio: A Photographic Record of Sculpture in the Making* (Oxford: Phaidon, 1980)

Elton, G.R., *Studies in Tudor and Stuart Politics and Government* (Cambridge: Cambridge University Press, 1983), vol 3

Elwin, M. (ed), *The Autobiography and Journals of Benjamin Robert Haydon* (London: Macdonald, 1950)

Engle, G., letter, *The Times*, 8 June 2002

Evans, K., letter, *The Times*, 2 January 2003

Ezard, J., review of *C.S. Lewis at the Breakfast Table*, in the *Guardian*, 21 August 1980

Fairhall, A.L., G.D. Lewen, W. Bialek and R.R. de van Steveninck, 'Efficiency and Ambiguity in an Adaptive Neural Code', *Nature*, CCCCXII, no. 6849 (2001)

Foot, P., *Natural Goodness* (Oxford: Clarendon Press, 2001)

Franklin, B., 'The Savages of North America', in J. Bigelow (ed), *Complete Works* (New York, 1887–88)

Fraser, J.T., *The Voices of Time* (London: Allen Lane, 1968)

Friedman, W.J., *About Time: Inventing the Fourth Dimension* (Cambridge, Mass.: MIT Press, 1990)

Gage, J., *Colour and Culture: Practice and Meaning from Antiquity to Abstraction* (London: Thames and Hudson, 1995)

—, *Colour and Meaning: Art, Science and Symbolism* (London: Thames & Hudson, 1999)

Gale, R.M., *The Philosophy of Time* (Sussex: Harvester Press, 1978)

Gamow, G., *My World Line: An Informal Biography* (New York: Viking Press, 1970)

Garfield, J.L., *The Fundamental Wisdom of the Middle Way: Nagarjuna's Mulamadhyamakarika* (New York: Oxford University Press, 1995)

Gaudapada, *Mandukyakarika*

Gay, V.P., 'Reductionism and Redundancy', *Zygon*, XIII (1978)

Gleick, J., *Genius: Richard Feynman and Modern Physics* (London: Little Brown, 1992)

Glucklich, A., *Sacred Pain: Hurting the Body for the Sake of the Soul* (Oxford: Oxford University Press, 2001)

Goldsmith, O., *The Citizen of the World, or Letters from a Chinese Philosopher Residing in London to his Friends in the Country* (London: Cooke, 2 vols, 1799)

Gouldner, A.W., *The Coming Crisis of Western Sociology* (London: Heinemann, 1971)

Grahame, K., *The Golden Age* (London: John Lane, 1915)

Grayson, C., *Leon Battista Alberti on Painting and on Sculpture* (London: Phaidon, 1972)

Hall, E., *Inventing the Barbarian* (Oxford: Oxford University Press, 1991)

Hamilton, J., *God, Guns and Israel* (Stroud: Sutton Publishing, 2004)

Hampshire, S., *Innocence and Experience* (London: Allen Lane, 1989)

Hanke, L., *Aristotle and the American Indians: A Study in Race Prejudice in the Modern World* (Chicago: Henry Regnery, 1959)

Hare, R.M., 'Ontology in Ethics', in T. Honderich (ed), *Morality and Objectivity: A Tribute to J.L. Mackie* (London: Routledge & Kegan Paul, 1985)

Harrod, R.F., *The Prof: A Personal Memoir of Lord Cherwell* (London: Macmillan, 1959)

Harvey, A., *Light upon Light: Inspirations from Rumi* (Berkeley: North Atlantic Books, 1996)

Hasan, A., *The Early Development of Islamic Jurisprudence* (Islamabad: Islamic Research Institute, 1970)

Hawking, S., *The Universe in a Nutshell* (London: Bantam Press, 2001)

Haydon, B., *The Autobiography and Journals of Benjamin Robert Haydon*, ed M. Elwin (London: Macdonald, 1950)

Heim, S.M., *Salvations: Truth and Difference in Religion* (Mayknoll: Orbis Books, 1995)

Henry, C.F.H. and W. S. Mooneyham, *One Race, One Gospel, One Task* (Minneapolis: World Wide Publications, 1967)

Henson, H., *Christian Morality* (Oxford: Clarendon Press, 1936)

—, *Bishoprick Papers* (London: Oxford University Press, 1946)

—, *Retrospect of an Unimportant Life* (Oxford: Oxford University Press, 1946)

Hewison, R., *Ruskin, Turner and the Pre-Raphaelites* (London: Tate Gallery Publishing, 2000)

Hick, J., *An Interpretation of Religion* (London: Macmillan, 1989)

Hirata Atsutane, *Kodo Taii*, in *Hirata Atsutane zenshu* (Tokyo, 1911)

Hoerl, C. and T. McCormack (eds), *Time and Memory: Issues in Philosophy and Psychology* (Oxford: Clarendon Press, 2001)

Honoré, D.D. (ed), *Trevor Huddleston: Essays on His Life and Work* (Oxford: Oxford University Press, 1988)

Honour, H. and J. Fleming, *A World History of Art* (London: Laurence King, 1999)

Hooker, R., *The Works ...* (Oxford: Clarendon Press, 1885)

Horace, *Epistolae*

Housman, A.E., *The Name and Nature of Poetry* (Cambridge: Cambridge University Press, 1962)

Hume, D., *A Treatise of Human Nature*, ed L.A. Selby-Bigge (Oxford: Oxford University Press, 1955)

—, An Enquiry Concerning the Principles of Morals, in R.P. Wolff (ed), The Essential David Hume (New York: Mentor Books, 1969)

Hunt, H.W., Pre-Raphaelitism and the Pre-Raphaelite Brotherhood, in D.S.R. Welland (ed), The Pre-Raphaelites in Literature and Art (London: Harrap, 1953)

Hutchings, E. (ed), "Surely You're Joking, Mr. Feynman!" Adventures of a Curious Character (Toronto: Bantam Books, 1986)

International Missionary Council, The Christian Message in Relation to Non-Christian Systems: Report of the Jerusalem Meeting of the International Missionary Council (London: Oxford University Press, 1928)

Iremonger, F.A., William Temple: His Life and Letters (London: Oxford University Press, 1949)

Jacob, F., The Possible and the Actual (Seattle: University of Washington Press, 1982)

Jenkins, M., The State Portrait, its Origin and Evolution (New York: College Art Association (Monographs on Archaeology and Fine Arts), iii, 1947)

John Paul II, Veritatis Splendor (London: Catholic Truth Society, 1993)

—, Crossing the Threshold of Hope (London: Jonathan Cape, 1994)

—, Letter of John Paul II to Women, Catholic Bishops' Conference, 1995

—, Evangelium Vitae (London: Catholic Truth Society, n.d.)

Johnson, C., The Language of Painting (Cambridge: Cambridge University Press, 1949)

Johnston, D. (ed), Faith-based Diplomacy: Trumping Realpolitik (Oxford: Oxford University Press, 2003)

—, 'Faith-based Diplomacy: Trumping Realpolitik', in Interreligious Insight, II (2004)

— and C. Sampson, Religion, The Missing Dimension of Statecraft (Oxford: Oxford University Press, 1994)

Joseph, R., The Naked Neuron: Evolution and the Languages of the Body and Brain (New York: Plenum, 1993)

Jowitt, K., New World Disorder: The Leninist Extinction (Berkeley: University of California Press, 1992)

Kant, I., Foundations of the Metaphysics of Morals, trans L.W. Beck (Indianapolis: Bobbs-Merrill, 1969)

Kemp, M. and M. Walker, Leonardo on Painting: An Anthology of Writings by Leonardo da Vinci with a Selection of Documents Relating to his Career as an Artist (London: Yale University Press, 1989)

Keown, D., The Nature of Buddhist Ethics (London: Macmillan, 1992)

Keynes, G., and B. Hill (eds), Samuel Butler's Notebooks (London: Jonathan Cape, 1951)

Kohak, E., The Embers and the Stars: A Philosophical Inquiry into the Moral Sense of Nature (Chicago: University of Chicago Press, 1984)

Krailsheimer, A.J. (trans), Pascal Pensées (London: Penguin Books, 1966)

Kroeber, A.L., 'The Eighteen Professions', American Anthropologist, XVII (1915)

—, *California Kinship Systems*, University of California Publications in American Archaeology and Ethology, XII (1917)

—, 'The Superorganic', *American Anthropologist*, XIX (1917)

—, 'History and Science in Anthropology', *American Anthropologist*, XXXVII (1935)

—, *The Nature of Culture* (Chicago: University of Chicago Press, 1952)

Kwok, D.W.Y., *Scientism in Chinese Thought, 1900–1950* (New Haven: Yale University Press, 1965)

Lambourne, L., *Victorian Painting* (London: Phaidon, 1999)

Lawrence, D.H., 'Art and Morality', in E.D. McDonald (ed), *Phoenix: The Posthumous Papers of D.H. Lawrence* (New York: Viking, 1972)

Leahy, M.P.T., *Against Liberation: Putting Animals in Perspective* (London: Routledge, 1991)

Lear, J., 'On Moral Objectivity', in S.C. Brown (ed), *Objectivity and Cultural Divergence* (Cambridge: Cambridge University Press, 1984)

LeDoux, J.E., *The Emotional Brain* (London: Weidenfeld, 1998)

—, 'Emotions: A View through the Brain', in R.J. Russell (ed), *Neuroscience and the Person* (Notre Dame: Notre Dame University Press, 1999)

Lee, K., *A New Basis for Moral Philosophy* (London: Routledge, 1985)

Lee, S.E. and W.K. Ho, *Chinese Art under the Mongols: The Yüan Dynasty (1279–1368)* (Cleveland: Cleveland Museum of Art, 1968)

Legge, J.G., *Rhyme and Revolution in Germany* (London: Constable, 1918)

Leitner, B., *The Architecture of Ludwig Wittgenstein* (New York: New York University Press, 1976)

Letwin, S.R., *The Pursuit of Certainty: David Hume, Jeremy Bentham, John Stuart Mill, Beatrice Webb* (Cambridge: Cambridge University Press, 1965)

Levi, P., *If This is a Man* (London: Penguin, 1979)

Lévi-Strauss, C., *The Savage Mind* (Chicago: Chicago University Press, 1967)

Leys, S., *The Analects of Confucius* (New York: W.W. Norton, 1997)

Lockhart, J.G., *Cosmo Gordon Lang* (London: Hodder & Stoughton, 1949)

Loehr, M., *The Great Painters of China* (Oxford: Phaidon, 1980)

Lonergan, B., *A Second Collection*, ed W.F.J. Ryan (Philadelphia: Westminster Press, 1974)

Loran, E., *Cézanne's Composition: Analysis of His Form with Diagrams and Photographs of His Motifs* (Berkeley: University of California Press, 1947)

Lowie, R., *History of Ethnological Theory* (New York: Farrer & Rinehart, 1937)

McGee, W.J., 'Some Principles of Nomenclature', *American Anthropologist*, VIII (1895)

Mackie, J.L., *Ethics: Inventing Right and Wrong* (London: Penguin, 1977)

Macksey, R. and E. Donato, *The Languages of Criticism and the Sciences of Man* (Baltimore: Johns Hopkins Press, 1970)

MacLaughlin, J., *Is One Religion as Good as Another?* (London: Burns & Oates, 1891)

Maclean, F., *A Person from England and Other Travellers* (London: Jonathan Cape, 1958)

MacLeish, A., *Streets in the Moon* (Boston: Houghton Mifflin, 1926)

McNaughton, D., *Moral Vision: An Introduction to Ethics* (Oxford: Blackwell, 1988)

Mandelbaum, D.G. (ed), *Selected Writings of Edward Sapir on Language, Culture, and Personality* (Berkeley: University of California Press, 1949)

Maritain, J., *The Range of Reason* (New York: Scribners, 1952)

Meerloo, J.A.M., 'The Time Sense in Psychiatry', in J.T. Fraser (ed), *The Voices of Time* (London: Allen Lane, 1968)

Meltzoff, A.N. and M.K. Moore, 'Imitation of Facial and Manual Gestures by Human Neonates', *Science*, CXCVIII (1977)

Mendelson, E. (ed), *W.H. Auden: Collected Poems* (London: Faber & Faber, 1994)

Miller, R.W., *Fact and Method: Explanation, Confirmation and Reality in the Natural and the Social Studies* (Princeton: Princeton University Press, 1987)

Moore, P., *E=mc²* (London: Quintet, 2002)

Moorehead, C., *Martha Gellhorn: A Life* (London: Chatto & Windus, 2003)

Mujeeb, M., *The Indian Muslims* (New Delhi: Munshiram Manoharlal, 1995)

Mumford, J., *The Question of Questions, Which, Rightly Resolved, Resolves All Our Questions in Religions* (London: Richards, 1841)

Murti, T.R.V., *The Central Philosophy of Buddhism: A Study of the Madhyamika System* (London: Allen & Unwin, 1955)

National Cyclopedia of Useful Knowledge (London: Charles Knight, 1848)

Nicholson, N., *The Lakers: The Adventures of the First Tourists* (London: Robert Hale, 1955)

Niebuhr, R., 'The Moral World of John Foster Dulles', *New Republic*, 1 December 1958

Nishijo. H. et al, 'Single Neuron Responses in Amygdala of Alert Monkey during Complex Sensory Stimulation with Affective Significance', *Journal of Neuroscience*, VIII (1988)

O'Connor, D.J., *Aquinas and Natural Law* (London: Macmillan, 1967)

O'Donohue, J., *Conamara Blues* (Sydney: Bantam, 2001)

O'Donovan, O.M.T., *Resurrection and Moral Order* (Leicester: Inter-Varsity Press, 1986)

Organ, T. W., *The Hindu Quest for the Perfection of Man* (Athens: Ohio University Press, 1970)

Orwell, G., *The Lion and the Unicorn* (London: Penguin 1982)

Osborne, H., *Theory of Beauty: An Introduction to Aesthetics* (London: Routledge & Kegan Paul, 1952)

Paige, K.E. and J.M. Paige, *The Politics of Reproductive Ritual* (Berkeley: University of California Press, 1981)

Palmer, H.P., *Joseph Wolff: His Romantic Life and Travels* (London: Heath Cranton, 1935)

Parsons, T. et al, *Theories of Society: Foundations of Modern Sociological Theory* (New York: The Free Press, 1965)

Paul VI, *Humanae Vitae* (Rome: Acta Apostolicae Sedis, 1968)

Paul, E.F. and J. Paul (eds), *Why Animal Experimentation Matters: The Use of Animals in Medical Research* (New Brunswick: Transaction Publishers, 2001)

Peierls, R., *Surprises in Theoretical Physics* (Princeton: Princeton University Press, 1979)

Pigden, C. (ed), *Russell on Ethics* (London: Routledge, 1999)

Poulet, G., *The Metamorphoses of the Circle*, trans C. Dawson and E. Coleman (Baltimore: Johns Hopkins Press, 1970)

Prestige, G. L., *The Life of Charles Gore* (London: Heinemann, 1935)

Race, A., *Christians and Religious Pluralism: Patterns in the Christian Theology of Religions* (London: SCM, 1983)

Rasa, A.E., C. Vogel and E. Voland, *The Sociobiology of Sexual and Reproductive Strategies* (London: Chapman and Hall, 1989)

Ratey, J. J., *A User's Guide to the Brain: Attention, Perception and the Four Theatres of the Brain* (New York: Little Brown, 2001)

Regan, T., *Defending Animal Rights* (Urbana: University of Illinois Press, 2001)

Reid, T., *The Works of Thomas Reid, D.D.* (Edinburgh, 1812)

Richards, I.A., *Poetry and Sciences* (New York: W.W. Norton, 1969)

Ricks, C. (ed), *The Poems of Tennyson* (London: Longmans, 1969)

Rolls, E.T., *The Brain and Emotion* (Oxford: Oxford University Press, 1999)

——, 'Neurophysiology and Functions of the Primate Amygdala, and the Neural Basis of Emotion', in J.P. Aggleton, *The Amygdala: A Functional Analysis* (Oxford: Oxford University Press, 2000)

Rosen, C., *The Classical Style: Haydn, Mozart, Beethoven* (London: Faber & Faber, 1971)

Rothenstein, J. (ed), *Stanley Spencer: The Man: Correspondence and Reminiscences* (London: Paul Elek, 1979)

Routh, G.V., *Towards the Twentieth Century* (Cambridge: Cambridge University Press, 1937)

Rucker, R., *Infinity and the Mind: The Science and Philosophy of the Infinite* (London: Paladin, 1984)

Russell, B., 'Reply to Criticisms', in P.A. Schlipp (ed), *The Philosophy of Bertrand Russell* (La Salle: Open Court Publishing Company, 1944)

——, *The Autobiography of Bertrand Russell* (London: Unwin Books, 1975)

Ryusaku Tsunoda (ed), *Sources of Japanese Tradition* (New York: Columbia University Press, 1964)

Sanday, W., *The Life of Christ in Recent Research* (Oxford: Clarendon Press, 1907)

Sapir, E., 'Do We Need a "Superorganic"?', *American Anthropologist*, XIX (1917)

——, 'The Status of Linguistics as a Science', *Language*, V (1929)

Saunders, N., *Divine Action and Modern Science* (Cambridge: Cambridge University Press, 2002)

Schacherl, L., 'Black Iris II', in J. Tesch and E. Hollmann (eds), *Icons of Art: The Twentieth Century* (Munich: Prestel, n.d.)

Scheler, M., *Schriften aus dem Nachlass, I, Zur Ethik und Erkenntnislehre* (Bern: Francke Verlag, 1957)

Schofield, M. et al (eds), *Doubt and Dogmatism: Studies in Hellenistic Epistemology* (Oxford: Clarendon Press, 1980)

Schomberg, J.D., *The Theocratic Philosophy of History; Being an Attempt to Impress upon History Its True Genius and Real Character; and to Present It, Not as a Disjointed Series of Facts, but as One Grand Whole, Exhibiting the Progress of the Social System, Tending under the Conduct of Its Divine Author, by Gradual and Almost Imperceptible Advances to Its Completion* (London: Whittaker and Co., 1839)

Scott, S.K. et al, 'Impaired Auditory Recognition of Fear and Anger Following Bilateral Amygdala Lesions', *Nature*, CCCLXXXV (1997)

Shelley, P.B., Letter to Maria Gladstone, 1 July 1820

Shweder, R.A. and R.A. LeVine, *Culture Theory: Essays on Mind, Self and Emotion* (Cambridge: Cambridge University Press, 1984)

Silvester, C., *The Literary Companion to Parliament* (London: Sinclair-Stevenson, 1996)

Singh, N.G.K., *The Name of My Beloved: Verses of the Sikh Gurus* (London: HarperCollins, 1995)

Sirén, O., *The Chinese on the Art of Painting* (Peking: Henri Vetch, 1936)

—, *Chinese Painting: Leading Masters and Principles* (London: IV, 1958)

Smith, W.J., in E.G. d'Aquili (ed), *The Spectrum of Ritual: A Biogenetic Structural Analysis* (New York: Columbia University Press, 1979)

Sommer, V., 'Infanticide among Free-Ranging Langurs *(Presbytis entellus)* at Jodhpur (Rajasthan/India): Recent Observations and a Reconsideration of Hypotheses', *Primates*, XXVIII (1987)

Spivey, N., *Enduring Creation: Art, Pain and Fortitude* (London: Thames & Hudson, 2001)

Sterne, L., *The Life and Opinions of Tristram Shandy* (London, 1948)

Stocking, G., 'From Physics to Ethnology: Franz Boas' Arctic Expedition as a Problem in the Historiography of the Behavioral Sciences', *Journal of the History of the Behavioral Sciences*, I (1965)

Stravinsky, I., *The Poetics of Music in the Form of Six Lessons* (Harvard: Harvard University Press, 1977)

Strong, R., *Portraits of Queen Elizabeth I* (Oxford: Clarendon Press, 1963)

Sullivan, M., *Symbols of Eternity: The Art of Landscape Painting in China* (Oxford: Clarendon Press, 1979)

Tacitus, *Annals*

Taylor, C., *Sources of the Self: The Making of Modern Identity* (Cambridge: Cambridge University Press, 1989)

Temple, W., *Readings in St. John's Gospel* (London: Macmillan, 1955)

Thatcher, M., Interview in *Women's Own*, 31 October 1987

Tilghman, B.R., *But Is It Art?* (Oxford: Blackwell, 1984)

Tippett, M., *Music of the Angels: Essays and Sketchbooks* (London: Eulenburg Books, 1980)

Tolkien, J.R.R., *The Lord of the Rings* (London: Folio Society, 1977)

Trinity College, Cambridge, 'The Portrait of Henry VIII', *Annual Record*, 2003

Tudge, C., *The Variety of Life: A Survey and Celebration of All the Creatures That Have Ever Lived* (Oxford: Oxford University Press, 2000)

Turner, J., *New Scientist*, 9 February 1984

Van Buitenen, *The Bhagavadgita in the Mahabharata: A Bilingual Edition* (Chicago: University of Chicago Press, 1981)

van der Ven, J.A., *Formation of the Moral Self* (Wheaton, Illinois: Eeerdmans, 1998)

Vasari, G., *Lives of the Painters, Sculptors and Architects*, trans G. du C. de Vere (London: Everyman's Library, 1996)

Verkamp, B.J., *The Indifferent Mean: Adiaforism in the English Reformation to 1554* (Athens: Ohio University Press, 1977)

Visiak, E.H. (ed), *Milton: Complete Poetry and Selected Prose* (London: Nonesuch Library, 1952)

Wagner, R., 'The Music of the Future', in *Art, Life and Theories* (New York, 1889)

Walker, R., *The Coherence Theory of Truth: Realism, Antirealism, Idealism* (London: Routledge, 1989)

Waller, J., *Fabulous Science: Fact and Fiction in the History of Scientific Discovery* (Oxford: Oxford University Press, 2002)

Walsh, H.W., *An Introduction to the Philosophy of History* (London: Hutchinson, 1951)

—, 'The Politics of Historical Interpretation', in *The Tropics of Discourse* (Baltimore: Johns Hopkins University Press, 1978)

Ward, L.F., review of G.H. Scribner, *Where Did Life Begin?*, in *American Anthropologist*, VI (1904)

Watson, B., *Chinese Rhyme-Prose: Poems in the Fu Form from the Han and Six Dynasties Periods* (New York: Columbia University Press, 1971)

Watts, F. (ed), *Science Meets Faith* (London: SPCK, 1998)

Waugh, E., *Decline and Fall*, ed D. Bradshaw (London: Penguin, 2000)

Weiskrantz, L., 'Behavioural Changes Associated with Ablation of the Amygdaloid Complex in Monkeys', *Journal of Comparative and Physiological Psychology*, XLIX (1956)

Welland, D.S.R. (ed), *The Pre-Raphaelites in Literature and Art* (London: Harrap, 1953)

Westley, D., *Morality and Its Beyond* (Mystic: Twenty-Third Publications, 1984)

Weyrauther, I., *Muttertag und Mutterkreuz: Der Kult um die 'deutsche Mutter' im Nationalsozialismus* (Frankfurt am Main, 1993)

Whorf, B.L., 'An American Indian Model of the Universe', in R.M. Gale (ed), *The Philosophy of Time* (Sussex: Harvester Press, 1978)

Wilde, O., *The Picture of Dorian Gray*, in *Complete Works of Oscar Wilde* (London: Collins, 1981)

Wilson, C.M. (Lord Moran), *Winston Churchill: The Struggle for Survival, 1940–1965* (London: Constable, 1966)

Wilson, F.A.W. and E.T. Rolls, 'The Primate Amygdala and Reinforcement: A Disassociation between Rule-based and Associatively Mediated Memory Revealed in Amygdala Neuronal Activity' (in preparation)

Wilson, P.W., *The Greville Diary* (London: Heinemann, 1927)

Wittgenstein, L., *Lectures and Conversations on Aesthetics, Psychology and Religious Belief*, ed C. Barrett (Oxford: Blackwell, 1966)

Wodehouse, P.G., *The Girl in the Boat* (London: Jenkins, 1956)

Wolff, R.P. (ed), *The Essential David Hume* (New York: New American Library, 1969)

Woodham-Smith, C., *The Reason Why?* (London: Constable, 1953)

Woods, G.F., *A Defence of Theological Ethics* (Cambridge: Cambridge University Press, 1966)

Wyschogrod, E., *Saints and Post-modernism: Revisioning Moral Philosophy* (Chicago: University of Chicago Press, 1990)

Yao, X., *Confucianism and Christianity: A Comparative Study of Jen and Agape* (Brighton: Sussex Academic Press, 1996)

Yee, C., *The Chinese Eye: An Interpretation of Chinese Painting* (London: Methuen, 1935)

Young, A., *Out of the World and Back* (London: Hart-Davis, 1961)

Zhang Yan-yuan, *Li-dai ming-hua-ji* (Taipei, 1971)

INDEX